Praise for *The Anxiety Healer's*

"*The Anxiety Healer's Guide for Clinicians* stands out as an exceptional resource that seamlessly integrates scientific research with actionable exercises. Its customizable interventions and practical strategies play a crucial role in empowering clients to manage and overcome anxiety effectively. I would highly recommend it to any therapist hoping to support their client in self-management of anxiety."

—**Israa Nasir, MHC-LP,** author of *Toxic Productivity* and creator of @well.guide

"Alison Seponara has crafted an exceptional clinical resource in *The Anxiety Healer's Guide for Clinicians*. With over 85 actionable strategies, this guide empowers clients to build their own anxiety toolkits, fostering a sense of ownership and collaboration in the therapeutic process. From cognitive behavioral techniques to guided imagery and other holistic tools, this book provides an array of effective psychoeducational as well as intervention-focused tools and worksheets that are both practical and transformative. It is an indispensable resource for any clinician dedicated to personalized, client-centered care."

—**Kate Truitt, PhD, MBA,** traumatic stress and resilience expert, psychologist, and author of *Keep Breathing* and *Healing in Your Hands*

"In our increasingly unstable and chaotic modern world, mental health practitioners must be well-equipped to address anxiety. *The Anxiety Healer's Guide for Clinicians* by Alison Seponara is an essential resource for therapists. I am thrilled to have this resource on my bookshelf! It offers an array of practical techniques and tools grounded in evidence-based research that therapists wcan readily use with their clients."

—**Liz Kelly, LICSW,** therapist and author of *This Book Is Cheaper Than Therapy: A No-Nonsense Guide to Improving Your Mental Health*

"Alison Seponara is an expert in the field, and she has the rare quality of practicing what she preaches. She has dedicated her life to helping clients and the masses through her social media presence, and this book will now allow her to have an even greater impact by providing other clinicians with the tools to help their own clients. As a fellow clinician, I'll absolutely be using this book as a guide!"

—**Sandi Christiansen, MS, LPC, CCTP, CCATP**

The Anxiety Healer's

GUIDE FOR CLINICIANS

Over 85 Cognitive Behavioral Strategies to Help Anxious Clients Calm the Mind and Body

Alison Seponara, MS, LPC
@theanxietyhealer

THE ANXIETY HEALER'S GUIDE FOR CLINICIANS
Copyright © 2024 by Alison Seponara

Published by
PESI Publishing, Inc.
3839 White Ave
Eau Claire, WI 54703

Cover and interior design by Emily Dyer
Editing by Jenessa Jackson, PhD

ISBN 9781683737810 (print)
ISBN 9781683737827 (ePUB)
ISBN 9781683737834 (ePDF)

All rights reserved.
Printed in the United States of America.

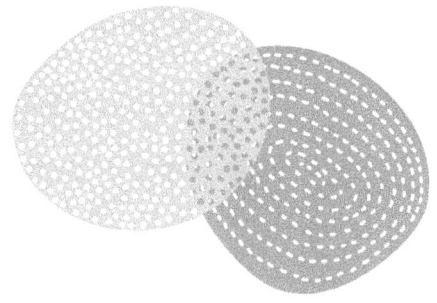

To all of the incredible healers in the world who have devoted their lives to helping others find internal peace and fighting to end the stigma of mental illness

Table of Contents

Introduction xi

PART I Laying the Groundwork 1

Chapter 1 • The Many Faces of Anxiety 3

Chapter 2 • The Neurophysiology of Anxiety 13

PART II Differential Diagnosis 27

Chapter 3 • The Classification of Anxiety Disorders 29

Chapter 4 • Understanding the Trauma Response 37

PART III Untangling Anxiety 49

Chapter 5 • The Anxiety Cycle 51

Chapter 6 • The Cognitive Behavioral Model 61

PART IV The Anxiety Healing Toolkit 77

Chapter 7 • Rewire the Anxious Brain 79

Chapter 8 • Breathwork and Grounding Techniques 103

Chapter 9 • Journaling, Affirmations, and Mirror Work 145

Chapter 10 • Visualization Strategies 165

Chapter 11 • Supplemental Holistic Healing Remedies 183

Chapter 12 • The Client Healing Toolkit 189

Chapter 13 • The Therapist Healing Toolkit 197

References 207

About the Author 211

List of Tools

THERAPEUTIC WORKSHEET •	My Life Experiences	4
PSYCHOEDUCATIONAL WORKSHEET •	Helpful Anxiety versus Harmful Anxiety	8
PSYCHOEDUCATIONAL HANDOUT •	Understanding Anxiety	11
PSYCHOEDUCATIONAL HANDOUT •	Autonomic Nervous System	16
PSYCHOEDUCATIONAL HANDOUT •	Tools to Increase Vagal Tone	19
THERAPEUTIC WORKSHEET •	My Physical Anxious Symptoms	22
THERAPEUTIC WORKSHEET •	My Anxious Body Clues	23
BETWEEN-SESSION WORKSHEET •	Daily Check-In: Physical Awareness	25
THERAPEUTIC WORKSHEET •	GAD-7 Anxiety	32
THERAPEUTIC WORKSHEET •	High-Functioning Anxiety	35
PSYCHOEDUCATIONAL WORKSHEET •	Perfectionism Looks Like . . .	36
THERAPEUTIC WORKSHEET •	Window of Tolerance Assessment	44
THERAPEUTIC WORKSHEET •	PTSD Checklist for *DSM-5*	46
THERAPEUTIC WORKSHEET •	The Anxiety Response	53
THERAPEUTIC WORKSHEET •	Safety Behaviors	55
BETWEEN-SESSION WORKSHEET •	Daily Check-In: Safety Behaviors	56
THERAPEUTIC WORKSHEET •	The Four Stages of Anxiety	58
PSYCHOEDUCATIONAL WORKSHEET •	The CBT Framework	63
THERAPEUTIC WORKSHEET •	My Core Beliefs	67
PSYCHOEDUCATIONAL HANDOUT •	Cognitive Distortions	69
THERAPEUTIC WORKSHEET •	Cognitive Distortions Checklist	71
THERAPEUTIC WORKSHEET •	Facts versus Opinions	74
IN-SESSION PRACTICE •	Brain Dump	80
BETWEEN-SESSION WORKSHEET •	Brain Dump	81
THERAPEUTIC WORKSHEET •	Feelings List	82
BETWEEN-SESSION WORKSHEET •	Daily Check-In: Emotional Awareness	83
IN-SESSION PRACTICE •	I-Statements	85
THERAPEUTIC WORKSHEET •	I-Statements	86
BETWEEN-SESSION WORKSHEET •	I-Statements	87
THERAPEUTIC WORKSHEET •	Identifying Triggers	89
BETWEEN-SESSION WORKSHEET •	Trigger Deep Dive	91
BETWEEN-SESSION WORKSHEET •	Daily Check-In: Triggers	93
IN-SESSION PRACTICE •	Rating Anxiety	95
THERAPEUTIC WORKSHEET •	The Anxiety Rating Scale	96
BETWEEN-SESSION WORKSHEET •	Daily Check-In: Rating Your Anxiety Level	98
PSYCHOEDUCATIONAL WORKSHEET •	Socratic Questioning	101
BETWEEN-SESSION WORKSHEET •	Thought Log	102

IN-SESSION PRACTICE •	Square Breathing	104
BETWEEN-SESSION WORKSHEET •	Daily Practice: Square Breathing	105
IN-SESSION PRACTICE •	Lion's Breath	107
BETWEEN-SESSION WORKSHEET •	Daily Practice: Lion's Breath	108
IN-SESSION PRACTICE •	Diaphragmatic (Belly) Breathing	110
BETWEEN-SESSION WORKSHEET •	Daily Practice: Diaphragmatic (Belly) Breathing	111
IN-SESSION PRACTICE •	Zen Word Breathing	112
THERAPEUTIC WORKSHEET •	My Zen Words	113
BETWEEN-SESSION WORKSHEET •	Daily Practice: Zen Word Breathing	114
IN-SESSION PRACTICE •	Bellows Breathing	116
BETWEEN-SESSION WORKSHEET •	Daily Practice: Bellows Breathing	117
IN-SESSION PRACTICE •	4-7-8 Breathing	119
BETWEEN-SESSION WORKSHEET •	Daily Practice: 4-7-8 Breathing	120
IN-SESSION PRACTICE •	Alternate Nostril Breathing	122
BETWEEN-SESSION WORKSHEET •	Daily Practice: Alternate Nostril Breathing	123
IN-SESSION PRACTICE •	Progressive Muscle Relaxation	126
BETWEEN-SESSION WORKSHEET •	Daily Practice: Progressive Muscle Relaxation	130
IN-SESSION PRACTICE •	You Are Safe Meditation	131
IN-SESSION PRACTICE •	Body Scan Meditation	133
IN-SESSION PRACTICE •	The Breath Meditation	135
IN-SESSION PRACTICE •	The Five Senses Meditation	137
BETWEEN-SESSION WORKSHEET •	The Five Senses Meditation	139
IN-SESSION PRACTICE •	Happy Place Meditation	141
BETWEEN-SESSION WORKSHEET •	Daily Practice: Grounding Meditations	143
BETWEEN-SESSION WORKSHEET •	Grounding Action Steps	144
PSYCHOEDUCATIONAL HANDOUT •	My Journaling Practice	146
THERAPEUTIC WORKSHEET •	Journal Practice	147
BETWEEN-SESSION WORKSHEET •	Self-Reflective Journal Prompts	148
BETWEEN-SESSION WORKSHEET •	Healing Journal Prompts	149
BETWEEN-SESSION WORKSHEET •	Self-Esteem Journal Prompts	150
BETWEEN-SESSION WORKSHEET •	Positive Mindset Daily Journal	151
PSYCHOEDUCATIONAL HANDOUT •	How to Create Affirmations	154
THERAPEUTIC WORKSHEET •	My Affirmation List	155
THERAPEUTIC WORKSHEET •	My Strengths	156
BETWEEN-SESSION WORKSHEET •	List of Affirmations	157
BETWEEN-SESSION WORKSHEET •	Five Things I Love About Myself	158
PSYCHOEDUCATIONAL HANDOUT •	Daily Affirmations Practice	159
BETWEEN-SESSION WORKSHEET •	Affirmation Log	160
IN-SESSION PRACTICE •	Introducing Mirror Work	162

THERAPEUTIC WORKSHEET • Mirror Work Reflection	163
BETWEEN-SESSION WORKSHEET • Mirror Work	164
THERAPEUTIC WORKSHEET • Creative Visualization	167
BETWEEN-SESSION WORKSHEET • Daily Visualization	168
THERAPEUTIC WORKSHEET • Future Self Visualization	169
BETWEEN-SESSION WORKSHEET • The Vision Board	171
IN-SESSION PRACTICE • The Clouds	172
IN-SESSION PRACTICE • The Pond	174
IN-SESSION PRACTICE • The White Sandy Beach	176
IN-SESSION PRACTICE • The Safe Place	178
IN-SESSION PRACTICE • The Happy Memory	180
BETWEEN-SESSION WORKSHEET • Daily Practice: Guided Imagery Meditation	182
PSYCHOEDUCATIONAL HANDOUT • Sample Anxiety Healing Toolkit	190
THERAPEUTIC WORKSHEET • Your Anxiety Healing Toolkit	194

Introduction

When I began my work as a therapist, I knew nothing about holistic psychology. All I really knew was that the brain is complex and that providing therapy is all about helping others to think differently. What I didn't realize is there's a whole aspect of the psychology field that I wasn't aware of—and I definitely wasn't taught in graduate school. This aspect considers the connection between the mind, body, and spirit.

As someone who has lived a life struggling with my own anxiety and panic attacks, I knew that healing had to encompass much more than what I had learned in graduate school. I started educating myself on Eastern perspectives on mental health and began looking for more holistic ways to calm myself and regulate my nervous system. In doing so, I found comfort in guided meditations, breathwork, and self-soothing exercises. Additionally, I found a therapist who had faith in my recovery—and shared my belief in holistic psychology. I didn't know it at the time, but I was in the process of creating my own anxiety healing toolkit while becoming a full-time licensed therapist. As I learned how to regulate my own anxiety, I began incorporating these holistic practices into my work with clients and found incredibly successful outcomes. As the years went by, I became more confident in my healing work and in my ability to heal myself and others, and I knew I needed to share my method with the world.

This inspired me to write *The Anxiety Healer's Guide* in 2021, in which I help readers create their own unique anxiety healing toolkit that will lead them on the road to recovery. Since the book's publication, I have heard from countless therapists who sought a similar clinically focused reference guide with healing tools and techniques they could use with anxious clients in session or as ongoing treatment exercises between sessions. A guide filled with practical and concrete techniques to help clients learn how to challenge intrusive thoughts while incorporating holistic tools to calm their minds and bodies. A guide that would help them train clients on how to create their own anxiety healing toolkit.

That is what I have created with *The Anxiety Healer's Guide for Clinicians*. This workbook provides a step-by-step guide that clinicians can use to help clients regulate their nervous systems and rewire their anxious minds. It is intended for clinicians who are interested in a more proactive, solution-focused approach to healing anxiety. Filled with evidence-based strategies from the field of cognitive behavioral therapy (CBT)—along with holistic tools that enhance the mind-body connection, like breathwork, guided imagery, and meditation—it will help clients build better emotional awareness, self-regulation, and cognitive processing. In addition, this book emphasizes the importance of a personalized approach to healing. It acknowledges that healing cannot be achieved through a one-size-fits-all solution; rather, it requires a tailored approach that considers the unique experiences and needs of each client. This framework draws on my expertise in working extensively with anxious clients for over 20 years, as well as my own experience recovering from anxiety since I was a child.

Unlike many other resources, this book goes beyond just providing theoretical knowledge and instead offers a practical, hands-on solution to help clients actively engage in their own healing process. This approach is crucial because it gives them a sense of ownership and control over their anxiety, making them feel they have a team alongside them as they work toward their well-being. For example, instead of merely explaining different anxiety disorders and their symptoms, this workbook encourages clients to identify their specific triggers and create personalized coping mechanisms. By giving them tangible strategies they can use in real-life situations, this workbook empowers clients to take charge of their anxiety rather than being passive recipients of treatment. This approach fosters a sense of collaboration between the client and clinician, making the healing process more effective and fulfilling.

What Is in This Book

To guide your clients in creating their own anxiety healing toolkit, you must first help them understand more about their own anxiety, which is why parts I through III of this workbook provide valuable education on how anxiety manifests and what it takes to heal. Included in these initial parts is information on the neurophysiology of anxiety, the different types of anxiety, the difference between anxiety and trauma, the anxiety cycle, and the CBT model. We know that psychoeducation is a key feature of treatment, and it's up to us as clinicians to provide anxious clients with information that can empower them to better understand their experiences and take an active role in their own healing journey. This awareness is fundamental to the healing process, as it allows individuals to make conscious choices about how to respond to anxiety-provoking situations.

By the time you get to part IV of this workbook, your client will be motivated and ready to create their healing toolkit! At this point, clinical practice may look more direct and action-oriented. The exercises in this part are meant to be used in session with your client to practice rewiring their anxious mind and self-soothing their anxious body in times of panic. Chapter 7, which covers rewiring of the anxious brain, is perhaps one of the most important chapters because this step takes *time and practice*! Remember that each client will work through their healing in a different manner. You may need to return to certain exercises in this book again and again, which is a *good* thing. This means your clients are doing the work.

Within each chapter, you will find the following tools:

- **Psychoeducational worksheets and handouts:** Since psychoeducation is an important aspect of anxiety healing, these tools educate clients about anxiety and its symptoms using clear and accessible language. This allows clients to see that anxiety is a natural response to stress and that it can be managed and overcome.

- **Therapeutic worksheets:** These in-session worksheets are designed to support clients in applying what they have learned and deepening their understanding of the concepts you cover together. By completing these worksheets, clients can identify their triggers, explore their thoughts and emotions, and develop coping strategies. This structured approach helps clients gain clarity and insight into their anxiety and provides them with useful tools to manage it effectively.

- **In-session practices:** These exercises will help clients learn new evidence-based healing skills, including CBT techniques, mindfulness practices, relaxation exercises, and other modalities.

- **Between-session worksheets:** These worksheets provide clients with an opportunity to apply the techniques they have learned in session to their daily lives. If clients are dedicated to reducing their anxiety, they must practice these tools every day—outside of session—even when their anxiety feels manageable. The more they practice these healing tools on a daily basis, the more they will notice a decrease in anxiety and an improvement in their overall mood.

By the completion of this workbook, your client will not only have a better understanding of their own anxious mind and body, but they also will have created their own unique toolkit of holistic practices to take with them when they leave your office. It is important to note that this workbook will be most effective for clients who have a higher level of self-awareness about their own limitations and are motivated to constantly maintain a lifestyle practice of mindfulness and healing. Remember that healing anxiety is a lifelong journey. As therapists, it's our duty to create a safe space for clients so they trust that we can help them get through their hardest moments in life. Doing so requires giving them education, guidance, and a collaborative experience while practicing hands-on, practical techniques. Healing also requires teaching them how to sit with discomfort, reframe their thought patterns, and regulate their nervous system.

Remember that all clients will experience and heal from anxiety differently. If the first tool from this book doesn't seem to help your client, or they still need a "boost," practice another one, and then another. Healing does not mean your clients will never feel anxious again, but with the holistic tools in this book, they will be able to feel self-sufficient throughout their anxious moments in life.

PART I
Laying the Groundwork

CHAPTER 1

The Many Faces of Anxiety

Anxiety is a complex and multifaceted condition that affects people in various ways. It can manifest differently from person to person, as some individuals may experience more intense physical symptoms (e.g., racing heart, sweating, shortness of breath), while others may experience more cognitive symptoms (e.g., excessive worry, negative thoughts, difficulty concentrating). The emotional symptoms can also vary, ranging from feeling restless or on edge to experiencing irritability or even panic attacks. Individuals may experience all of these symptoms or only some of them. What's more is that the way we take in information, interpret bodily sensations, and relate to the world is unique, so our interpretation and description of anxiety symptoms may be subjective (Novak, 2021).

So why do people experience anxiety in such different ways? It all has to do with individual differences. Because we all have varying life experiences, anxiety can look very different in terms of its intensity and diagnostic characteristics (Novak, 2021). For example, a client who experienced a traumatic event in their past may develop anxiety as a result of that experience, such as someone who served in combat and has a posttraumatic stress disorder (PTSD) diagnosis. On the other hand, a client who grew up with an absent father may develop attachment issues and have a generalized anxiety disorder (GAD) diagnosis. It's essential to recognize that these fears can be deeply rooted and may require individualized approaches to healing. The following worksheet will help give you more insight into the life experiences that may have contributed to your client's present maladaptive behaviors. For a more thorough trauma assessment, refer to chapter 4.

THERAPEUTIC WORKSHEET

My Life Experiences

Share more about your life experiences by answering the following questions. When you're finished, discuss your answers with your therapist.

Did you enjoy your childhood? Why or why not?

_____
_____
_____
_____

How was your school experience?

_____
_____
_____
_____

Did you have secrets in your household? If so, what kinds?

_____
_____
_____
_____

What type of friendships did you have?

_____
_____
_____
_____

Did anyone close to you die? If so, who?

_____
_____
_____
_____

What were holidays like in your household?

_____
_____
_____
_____

How was discipline handled in your family?

_____
_____
_____
_____

Anxiety's Purpose: When It's Helpful versus Harmful

When Anxiety Is Helpful

Contrary to popular belief, anxiety can actually be helpful in certain situations. Our bodies are biologically wired to experience anxiety as a response to perceived threats or dangers. This natural response triggers a series of physiological changes that prepare us to either confront the perceived threat or flee from it, which is often referred to as the fight-or-flight response. To help a client understand this, you might ask them to imagine that they are walking alone at night when they suddenly hear footsteps approaching quickly from behind. In this scenario, anxiety kicks in, alerting the client to a potential threat and prompting their body to release stress hormones, which may increase their heart rate and sharpen their senses. This heightened state of awareness can provide the client with the necessary adrenaline and energy to protect themselves or escape from danger (Novak, 2021).

Evolutionary psychologists believe that back in the prehistoric days, anxiety was influenced by natural selection, meaning that it got stronger because it kept humans alive; it helped them to immediately recognize signs of danger, ignite their fight-or-flight response, and make quicker decisions (Novak, 2021). In our current world, most situations that trigger anxious feelings are not life-threatening, but moderate amounts of short-term anxiety can continue to serve a useful purpose by motivating people to solve problems and achieve goals—even those that may not sound pleasurable (Strack et al., 2017). That's because anxiety triggers the release of adrenaline, which increases heart rate, promotes blood flow to the brain, and produces a consequent rush of oxygen that collectively forces us to concentrate on the problem at hand and cope with it in a constructive manner. In this way, anxiety can help people accomplish things like completing schoolwork, fulfilling work duties, or doing household chores. It can also motivate people to prepare for nerve-racking situations, such as giving a presentation, taking a test, or having a hard conversation. Anxiety really drives us to do things in a way that few other feelings do (Meek, 2021), but it can be harmful if it becomes too consuming and difficult to cope with.

When Anxiety Is Harmful

As explained in the previous section, a small amount of anxiety can help people focus on completing goals on a short-term basis. However, long-term, chronic anxiety can have a detrimental effect on the mind and body. Someone with severe anxiety may be preoccupied with negative thoughts and unable to relax, leading to feelings of restlessness or irritability and difficulty concentrating. Chronic anxiety affects the physical body, too, as long-term exposure to stress hormones can weaken the immune system and make clients more susceptible to illnesses. When a client is stuck in a state of high alert, their heart rate increases, their breathing becomes shallow, and stress hormones flood their system. This stress response can have severe consequences, including muscle tension, respiratory problems, headaches, heart disease, memory loss, and digestive issues (Medanta, 2019).

Imagine a scenario where a client constantly worries about what others think of them. Physiologically speaking, their body perceives this situation as a threat, and even though they are not in any imminent danger, it activates the same stress response system as if they were. Over time, this heightened state of arousal can wear down the body and have significant consequences on their overall health. Therefore, it is important for clients to recognize the differences between helpful and harmful anxiety so they can take steps to manage and reduce the latter. Understanding the impact of long-term, chronic anxiety on the body and implementing stress-reducing strategies are essential in achieving freedom from anxiety and improving one's overall quality of life.

PSYCHOEDUCATIONAL WORKSHEET

Helpful Anxiety versus Harmful Anxiety

Anxiety can be beneficial, as it lets us know when we are in danger and helps us to complete tasks. However, long-term, chronic anxiety can have a detrimental effect on the mind and body. Review the following table to identify the differences between *helpful* anxiety and *harmful* anxiety. Then, consider when your own anxiety has been helpful and harmful.

Helpful Anxiety	**Harmful Anxiety**
Shows up when there's a true threat	Shows up when there's no true threat
Proportional to the issue	Is excessive compared to the issue
Appears when there's an action to take and time to take it	Shows up when you are feeling scared or worried and hinders you from taking action by sending you into a fight, flight, or freeze state
Goes away when the situation has passed	Does not go away, even after the situation has passed
Is driven by your values, making you more like the person you want to be	Takes you away from your values, making you less like the person you want to be

When, where, and how has your anxiety been *helpful*?

_____
_____
_____
_____

When, where, and how has your anxiety been *harmful*?

_____
_____
_____
_____

Everyday Stress versus Anxiety Disorder

Anxiety is a buzzword these days, and for good reason—most clients either have anxiety or know someone who struggles with it. Although anxiety is a normal response to many different situations in life, for some clients, anxiety can become so overwhelming and unmanageable that it starts to affect their quality of life. When this occurs, it is a sign that an anxiety disorder is present. Clients with anxiety disorders may experience several uncomfortable physical sensations and health problems while also feeling trapped by thoughts and feelings that never seem to subside. These anxious thought patterns are continuous, leading many clients to constantly feel on edge. Depending on their specific symptoms, the diagnosis may differ from one client to the next, but this "quality of life" indicator is a major difference between everyday stress and an anxiety disorder.

Indeed, as specified in the *Diagnostic and Statistical Manual for Mental Disorders* (*DSM-5*; American Psychiatric Association, 2013), one of the primary considerations in determining whether a client meets diagnostic criteria for an anxiety disorder is whether the intensity and duration of their symptoms interfere with daily functioning for a prolonged period of time (typically at least six months). Not only must their symptoms be persistent across time, but the symptoms must interfere with daily activities like job performance or relationships—in other words, these are activities that affect the client's quality of life.

However, there are also many clients with anxiety who experience bothersome symptoms but don't meet the threshold for a diagnosable disorder. When this occurs, a client is said to have subclinical anxiety. Many therapists still consider subclinical anxiety to be a disorder if symptoms are "clinically significant." This means that clients' symptoms (Zolfagharifard, 2023):

- Cause distress (are upsetting and distracting on a near-daily basis)
- Impair daily functioning (result in problems such as lack of sleep, loss of appetite, reduced productivity, or relationship tension)
- Are constant over time (show up regularly for weeks or months)

Remember that anxiety isn't always a bad thing. Clients need the adrenaline rush that anxiety provides to stay alert or propel themselves to action. But when anxiety is constant or overwhelming and also interferes with daily life, they may have an anxiety disorder. Ask your clients the following questions to assess whether they are experiencing everyday stress or severe anxiety that may reach diagnostic threshold:

1. Are you constantly tense, worried, or on edge?
2. Does your anxiety interfere with your work, school, or family responsibilities?

3. Do you believe that something bad will happen if certain things don't happen in a certain way?
4. Do you avoid everyday situations or activities that cause you anxiety?
5. Do you experience sudden or unexpected attacks of heart-pounding panic?

In chapter 3, we will take a deeper look into the classification of anxiety disorders. In the meantime, you can use the following handout to educate your clients about anxiety. Whether a diagnosis is present or not, this workbook will provide your client with the healing tools needed to live a productive life despite their anxiety.

PSYCHOEDUCATIONAL HANDOUT

Understanding Anxiety

What Is Anxiety?

Anxiety is a normal human emotion characterized by fear, worry, or uncertainty. Most of us stress on a regular basis about things like money, work, school, relationships, and family, but afterward we usually calm down and feel better. But if someone is struggling with severe anxiety, their feelings of fear and worry never seem to subside. They may constantly feel on edge and catastrophize about everyday situations—always thinking of the worst-case scenario. Most times these thoughts are unrealistic, but the brain tricks us into believing that they're true, which only creates more fear and less rational thinking patterns. When the brain has these thoughts, it feels nearly impossible to think about anything else. The body then responds with any of the following symptoms: racing heart, upset stomach, headache, muscle tightness, shortness of breath, or any number of other physical symptoms that may feel debilitating in that moment.

When Does Anxiety Become a Disorder?

Anxiety falls along a spectrum, ranging from occasionally helpful to chronic and severe—meeting criteria for a diagnosable disorder. It is also possible for someone to experience unhelpful anxiety but not meet the criteria for an anxiety disorder, which begs the question: When does anxiety cross the threshold and meet diagnostic criteria for a disorder? One of the primary considerations in determining an anxiety disorder is whether the intensity and duration of the person's symptoms interfere with their day-to-day functioning for a prolonged period of time (typically at least six months). To determine whether your anxiety is more than just everyday stress, here are some questions to ask yourself:

1. Are you constantly tense, worried, or on edge?

2. Does your anxiety interfere with your work, school, or family responsibilities?

3. Do you believe that something bad will happen if certain things don't happen in a certain way?

4. Do you avoid everyday situations or activities that cause you anxiety?

5. Do you experience sudden or unexpected attacks of heart-pounding panic?

CHAPTER 2

The Neurophysiology of Anxiety

Mental illness is a complex and multifaceted phenomenon that has long been misunderstood. However, as scientists delve deeper into the workings of the human brain, they are discovering that mental illness is a result of many different factors. Most mental illnesses are not caused by just one thing—it's often a combination of factors like genetics, environment, trauma, lifestyle habits, and brain chemistry. This shift in understanding has been driven by advancements in the field of neuroscience, which have provided us with a deeper understanding of the biological mechanisms underlying mental illness and shed light on the different brain patterns among those who struggle with such conditions. In particular, changes in brain structure, chemistry, and function have been observed among individuals with mental disorders, providing evidence that mental illness is not a result of personal weakness or character flaws. By recognizing these biological underpinnings, we can foster a more compassionate and informed society that supports clients in their journey toward healing and recovery. Let's take a look at which parts of the brain contribute to the stress response.

The Limbic System

Anxiety is a complex emotion that originates in the brain, and anxiety disorders are thought to result from a disruption in the limbic system. In particular, scientists have found that individuals with anxiety disorders have more activity in the limbic system (Northwestern Medicine, 2020), which is known as the brain's emotional processing center. The limbic system is composed of the hippocampus, amygdala, hypothalamus, and thalamus.

The amygdala is a small, almond-shaped structure that plays a key role in regulating fear and anxiety. When someone encounters a potentially threatening or dangerous situation, the amygdala gathers incoming sensory information from the environment (e.g., through visual cues or auditory signals) to determine whether the situation is a cause for concern. For example, if a client sees a snake slithering toward them, this visual information is transmitted to the amygdala, which recognizes the danger and quickly sends a distress signal to the hypothalamus.

The hypothalamus, in turn, activates the body's autonomic nervous system, which is a network of nerves that is responsible for regulating involuntary body functions like breathing, blood pressure, and heartbeat.

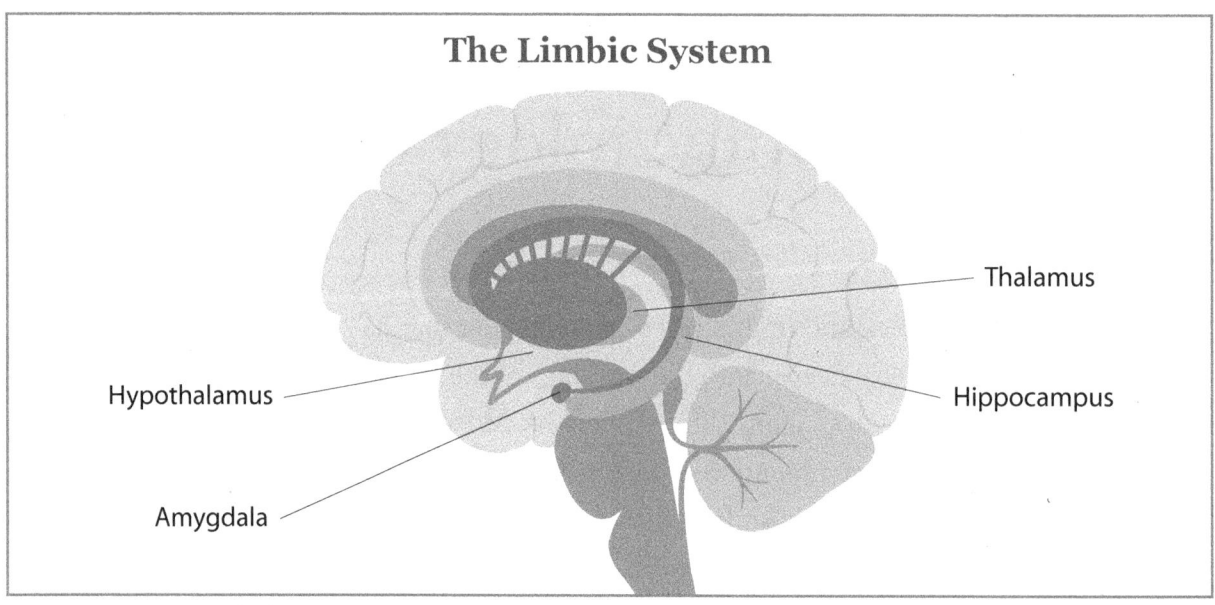

The Autonomic Nervous System

The autonomic nervous system is made up of two components: the sympathetic nervous system and the parasympathetic nervous system. The *sympathetic nervous system* is often likened to a gas pedal in a car. When a client perceives danger or a threat, this system is activated, triggering what is commonly known as the fight-or-flight response. In this state, the body is flooded with stress hormones like adrenaline and cortisol, providing the client with a burst of energy and heightened alertness that allows them to quickly respond to the threat at hand. These hormones increase heart rate, blood pressure, and breathing rate, which supplies more oxygen and nutrients to the muscles and brain, allowing the muscles to react more powerfully and move faster. It's important to note that individuals enter this fight-or-flight mode subconsciously, and the body's reaction to a threat is instinctual and involuntary (McCorry, 2007). In other words, it's not a choice people make.

The problem is, in today's world, the body doesn't know the difference between real and perceived danger, which means that the anxious brain can perceive non-threatening situations as dangerous. For example, imagine that a client comes to see you and reports that he hasn't been able to find a job in a year because he has severe anxiety during job interviews. He experiences a racing heart, nausea, shortness of breath, dizziness, and brain fog. Although his sympathetic nervous system has been activated and he is experiencing that same fight-or-flight response as our primal ancestors, there is no real danger. One of the ways you can help this client is by teaching him physiological regulation exercises to activate his parasympathetic nervous system, which we will discuss next.

In contrast to the sympathetic nervous system, the *parasympathetic nervous system* is often likened to a car's brakes, as it works to slow clients down and return the body to baseline once the perceived stressor has passed. When the parasympathetic nervous system is activated, metabolism decreases, heart rate slows, muscles relax, breathing becomes slower, and even blood pressure decreases. Because the parasympathetic nervous system calms the body down and promotes growth and repair, it is often known as the rest-and-digest response. The best way to activate the parasympathetic nervous system is through strategies that stimulate the vagus nerve, which functions as the communication highway between the brain and the body. Its main function is to tell the body when it's time to relax and de-stress, but sometimes it needs to be stimulated in order to see long-term improvements in mood, well-being, and resilience. You will learn more about the vagus nerve in the next section when we discuss polyvagal theory.

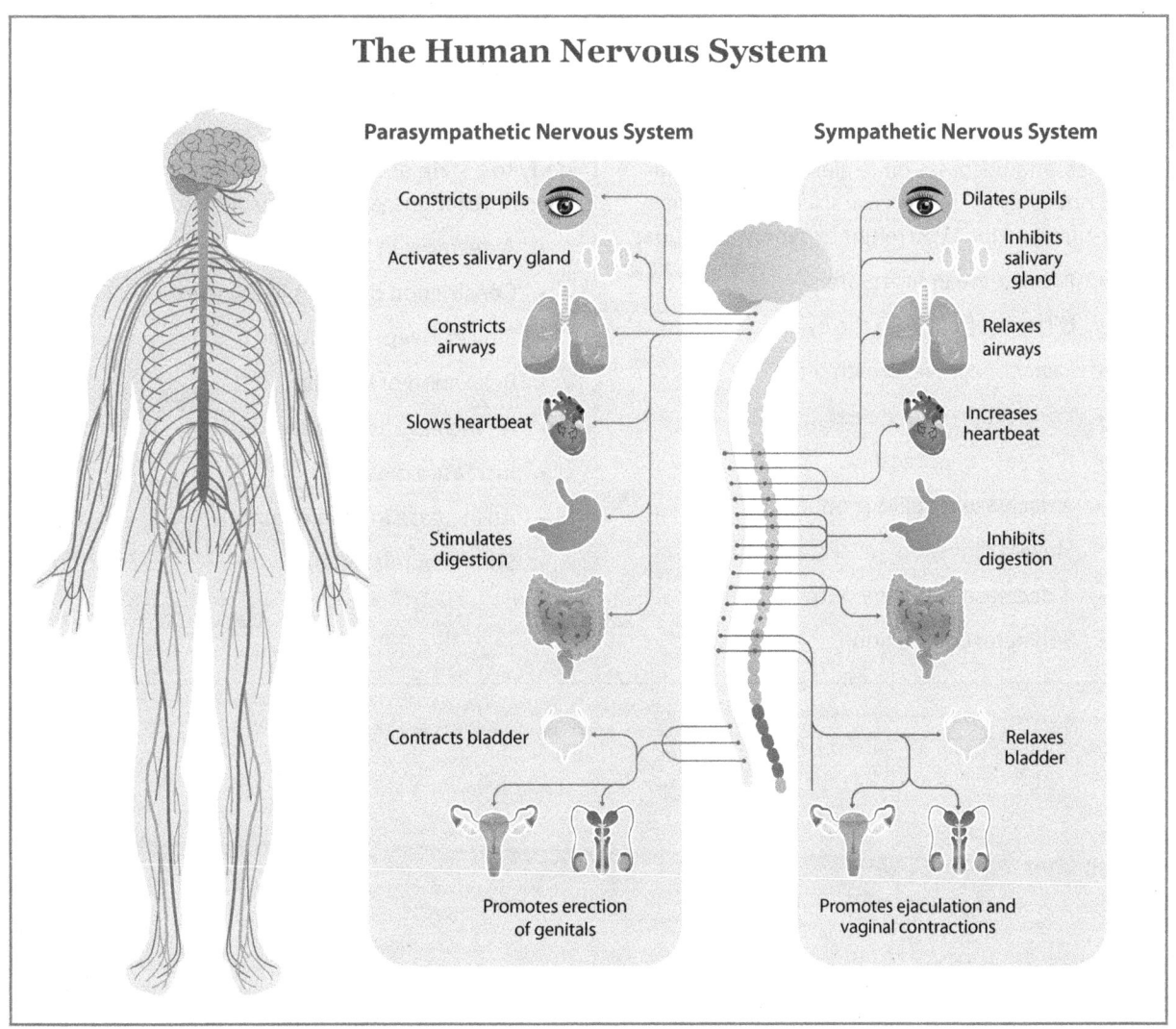

PSYCHOEDUCATIONAL HANDOUT

Autonomic Nervous System

The autonomic nervous system (ANS) regulates many of your internal organs and is made up of two components:

1. Sympathetic nervous system (SNS): Known as your fight-or-flight system, it is activated when you are anxious.

2. Parasympathetic nervous system (PNS): Known as your rest-and-digest system, it is activated when you are calm.

The following table shows how the body physically responds when each component of the ANS is activated. Do you notice any of these changes to your body when you're anxious?

Sympathetic Nervous System	Parasympathetic Nervous System
When the SNS is activated, the body speeds up, tenses, and becomes more alert. Functions that are not essential for survival shut down. Some specific reactions you might experience include: • An increase in heart rate • Dilation of pupils in the eyes • Faster, shallower breathing • Contraction of muscles • A release of adrenaline • A decrease in saliva production • Slowed digestion • A decrease in urinary output • Sphincter contraction	The PNS counterbalances the SNS. It restores the body to a state of calm or homeostasis. Some specific reactions you might experience include: • A decrease in heart rate • Constriction of pupils in the eyes • Slower, deeper breathing • Relaxation of muscles • An increase in saliva production • Increased digestion • An increase in urinary output • Sphincter relaxation

Polyvagal Theory

As we move through life, we all experience moments when we feel safe and others when we feel discomfort or danger. These ups and downs are a natural part of the human experience—just like the range of emotions that come with it. To better understand what is happening in the brain and body during these experiences, we can turn to Dr. Stephen Porges's polyvagal theory, which provides an insightful framework that sheds light on the complex workings of the autonomic nervous system. According to polyvagal theory, the autonomic nervous system is organized into a three-part structure that responds to cues of danger or safety in a hierarchical fashion: (1) the dorsal vagal system, (2) the sympathetic nervous system, and (3) the ventral vagal system (Porges, 2007). Understanding this hierarchy can help your clients better comprehend their physiological reactions to stress and anxiety.

The first stage in the hierarchy is the dorsal vagal system, which is the oldest part of the parasympathetic nervous system and is in charge of the freeze or immobilization response. When faced with an overwhelming threat, the human body has the innate capacity to shut down. This response can be seen in animals playing dead or in humans who become frozen in an extreme state of fear. When clients immobilize themselves, they attempt to escape the immediate danger by appearing harmless or unnoticed. While this response can be adaptive in certain situations, persistent freeze responses can contribute to chronic anxiety.

The second stage in the evolutionary hierarchy is the sympathetic nervous system, which, as discussed earlier, is in charge of the fight-or-flight response. When an individual is confronted with danger, their body undergoes a cascade of physiological changes that prepare them to either fight off the threat or flee from it. This response can be lifesaving in acute situations, but like continual freeze responses, prolonged activation of the sympathetic nervous system can lead to chronic anxiety, wear down the body, and negatively impact overall well-being.

The newest addition to the hierarchy of responses is the ventral vagal system, which is the part of the parasympathetic nervous system that responds to feelings of safety and connection. When a client's ventral vagal pathway is activated, they experience a sense of well-being. They are able to connect with others, form meaningful relationships, and experience a deeper sense of belonging. Social engagement is not just about having conversations or being in the presence of others; it is about feeling seen, heard, and understood (Porges, 2009). Co-regulation is a crucial component of the social engagement system, and it occurs when the nervous systems of two or more individuals interact in a way that supports mutual well-being and emotional connection. This can lead to increased feelings of calmness, relaxation, and emotional well-being (Porges, 2007).

According to polyvagal theory, our neural circuits are constantly scanning the cues in our environment and categorizing them as safe, dangerous, or potentially life-threatening so we can determine how to move through this hierarchical system. This process is known as *neuroception,* and it occurs automatically and operates at an unconscious level, meaning that we are not aware of it happening (Porges, 2004). Through neuroception, our nervous system is constantly on high alert, searching for cues and signals that might indicate potential danger. This can include nonverbal cues such as body language, facial expressions, tone of voice, and even the overall energy of the environment. Our brain processes this information at an unconscious level and then guides our behavioral and physiological responses accordingly (Clarke, 2023).

Neuroception is a process that occurs within our bodies, specifically through our vagus nerve. As discussed earlier, the vagus nerve plays a vital role in regulating the body's autonomic functions such as heart rate, breathing, and digestion. The vagus nerve connects the brain to several key organs, including the heart, lungs, and digestive system, making it the longest cranial nerve in the body. For this reason, it is also known as the *wandering nerve.* One way to measure the health of a client's nervous system is by assessing their vagal tone, which reflects the extent to which their body can return to baseline after stress. People with *high vagal tone* are able to bounce back more quickly and return to a state of equilibrium. On the other hand, people with *low vagal tone* have an overactive stress response system, leading to chronic anxiety symptoms and difficulty managing stress (Porges, 2007). The following handout contains strategies a client can use to increase their vagal tone, which will help to reduce inflammation in their body and better regulate their stress response.

PSYCHOEDUCATIONAL HANDOUT

Tools to Increase Vagal Tone

Massage

Research shows that light to moderate massages on areas of the body near the vagus nerve can increase vagal tone and slow heart rate. This includes massaging your feet as well as the right side of your throat (Lu et al., 2011).

Cold Exposure

Acute cold exposure has been shown to activate the neurons that are part of the vagal nerve pathway. Exposing yourself to cold on a regular basis can also lower your sympathetic nervous system's fight-or-flight response and increase parasympathetic activity through the vagus nerve (Jungmann et al., 2018). Here are some ideas to get you started:

- Take a cold shower. You can start off slow by finishing your next shower with at least 30 seconds of cold water and see how you feel. Work your way up to longer periods of time.
- Try a cold plunge in a bath of ice and water.
- Go outside in cold temperatures with minimal clothes.
- Submerge your face in ice-cold water.

Breathwork

Deep and slow breathing is another way to stimulate your vagus nerve. Practice slowing down your breath by taking about six breaths over the course of a minute. Make sure you breathe in deeply from your diaphragm. When you do this, your stomach should expand outward as you inhale and then fall back down as you take a long and slow exhale. This is key to stimulating the vagus nerve and reaching a state of relaxation (Gerritsen & Band, 2018).

Meditation

Meditation may sound intimidating if you've never tried it before, but it is one of the most effective healing techniques to stimulate the vagus nerve and increase vagal tone. You can start with just a few minutes a day. There are also apps like *Headspace* and *Calm* that can provide you with guided sessions if you would benefit from more structure.

Exercise

While it's no surprise that exercise is good for the mind and body, physical activity has also been shown to stimulate the vagus nerve. This may explain why exercise has such beneficial effects on the brain and overall mental health (Kai et al., 2016). You should aim for 30 to 60 minutes of exercise every day, whether it's walking, lifting weights, doing yoga or Pilates, going to a workout class, playing sports, or dancing.

Singing, Humming, Chanting, and Gargling

The vagus nerve is connected to the muscles in the back of your throat, so doing activities that engage the vocal cords—like singing, humming, chanting, and gargling—are all ways to activate your vagus nerve. One easy technique is to inhale deeply, and as you exhale, make a *hmmm* sound through closed lips, feeling the vibration travel down your throat.

Socializing and Laughing

Socializing and laughing can reduce your body's stress hormones, not only when *you* laugh but even when you hear laughter (Fujiwara & Okamura, 2018). Laughter sends a signal to your body to relax and stimulates the vagus nerve, so make it a point to hang out and laugh with your friends as much as possible.

Probiotics

It's becoming increasingly clear that the bacteria in your gut affect brain function via the vagus nerve. Because of this gut-brain connection, you can improve vagal tone by taking probiotics (Appleton, 2018). There are two specific strains that are directly related to the gut-brain connection as it relates to mood, anxiety, and depression: Lactobacillus and Bifidobacterium.

Physical Manifestations of Anxiety

Since the mind and body are connected, anxiety can create a host of physiological symptoms that set off a chain reaction in the body. Earlier, we discussed how anxiety can result in increased heart rate, elevations in blood pressure, and tense muscles. Other physical symptoms that clients may experience include shaking, hot flashes, chills, tingly arms and legs, fatigue, nausea, dizziness, and headaches. These physical responses can feel quite frightening when they happen—especially for clients who experience intense and severe panic attacks—but they are not dangerous. (However, you should confirm that your client has ruled out any other health conditions that may be causing these symptoms.)

It is important for clients to learn about their body cues and to become more aware of how their body responds to anxious thinking patterns. Building this physical awareness not only normalizes the somatic experience of anxiety, but also allows clients to identify the tools they can use to physiologically regulate the body. The following worksheets can help your clients gain a better understanding of their body's response to anxiety.

THERAPEUTIC WORKSHEET

My Physical Anxious Symptoms

Place a check mark by the most common physical symptoms you experience when anxiety strikes. Of the symptoms you mark, note which three happen the most frequently.

- ☐ Increased heart rate
- ☐ Shortness of breath
- ☐ Sweating
- ☐ Shaking
- ☐ Nausea
- ☐ Hyperventilation
- ☐ Chest pain
- ☐ Lightheadedness
- ☐ Fainting
- ☐ Muscle weakness
- ☐ Sweating
- ☐ Feeling of choking or throat tightening

- ☐ Muscle tension (e.g., clenching your jaw)
- ☐ Feeling weak or tired
- ☐ Hot flashes
- ☐ Chills
- ☐ Dry mouth
- ☐ Headache
- ☐ Gastrointestinal symptoms (e.g., nausea, cramping, diarrhea)
- ☐ Increased urination frequency
- ☐ Other: _____

My top three most common physical symptoms are:

1. _____

2. _____

3. _____

THERAPEUTIC WORKSHEET

My Anxious Body Clues

Color in the parts of the body where you experience physical sensations when you are anxious. Then write a detailed, specific description of these sensations, as well as what you are thinking when you feel them.

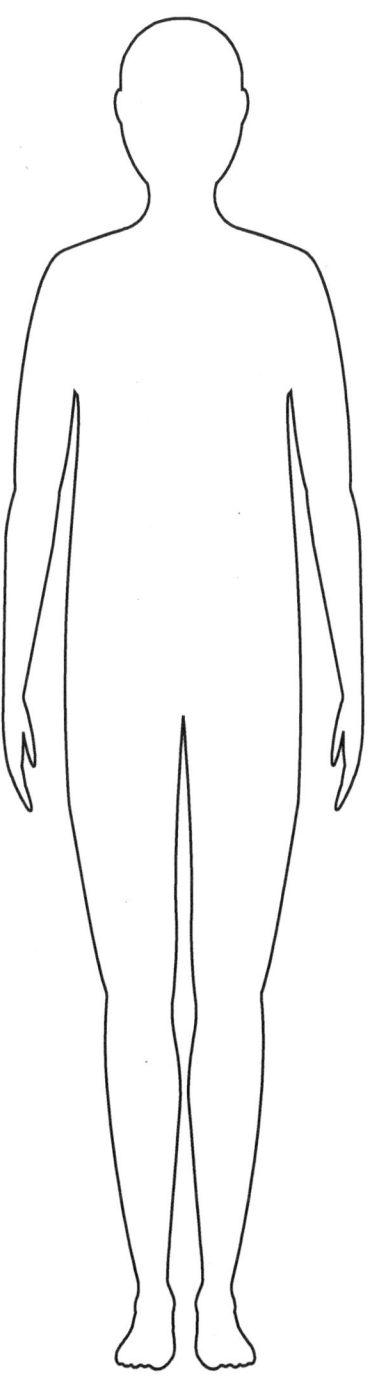

Description 1

I feel anxious in my: _____

The sensations I am feeling are: _____
_____

When I feel this way, I am thinking: _____
_____
_____

Description 2

I feel anxious in my: _____

The sensations I am feeling are: _____
_____

When I feel this way, I am thinking: _____
_____
_____

Description 3

I feel anxious in my: _____

The sensations I am feeling are: _____
_____

When I feel this way, I am thinking: _____
_____
_____

BETWEEN-SESSION WORKSHEET

Daily Check-In: Physical Awareness

Complete this worksheet every day to become familiar with the patterns of your physical anxiety symptoms. If needed, you can use the checklist from the *My Physical Anxious Symptoms* worksheet to reference your previous symptoms. Under the Reflection section for each day, describe your day, including the things you did, the places you went, or the people you interacted with.

Day/Time	Physical Symptoms	Intensity (1–10)
Monday		
Reflection:		
Tuesday		
Reflection:		
Wednesday		
Reflection:		

Day/Time	Physical Symptoms	Intensity (1–10)
Thursday		
Reflection:		
Friday		
Reflection:		
Saturday		
Reflection:		
Sunday		
Reflection:		

PART II
Differential Diagnosis

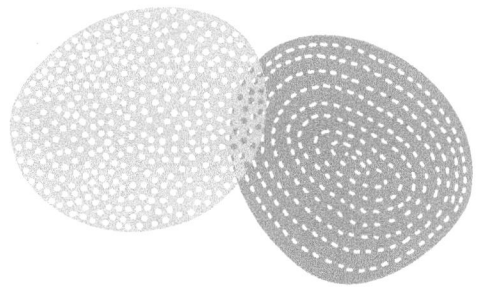

CHAPTER 3

The Classification of Anxiety Disorders

Over the past decade, there has been quite a change in the classification of anxiety disorders in the *Diagnostic and Statistical Manual of Mental Disorders*, with the *DSM-5* providing a more comprehensive and nuanced understanding of these disorders. For example, GAD is now recognized as a distinct disorder, whereas it was previously grouped with other anxiety disorders like social anxiety disorder and panic disorder. The separate classification of GAD acknowledges that individuals with this disorder experience excessive worry and anxiety across various aspects of their lives, rather than specific situations or objects. This change in classification allows for a more targeted and effective treatment approach.

Another change is the inclusion of separation anxiety disorder (SAD) as a stand-alone diagnosis. SAD was previously considered a disorder only affecting children but has now been recognized as a condition that can persist into adulthood. This change allows for better identification and treatment of individuals who experience significant distress and anxiety when separated from attachment figures.

Other notable changes include the removal of obsessive-compulsive disorder (OCD) and PTSD from the broader anxiety disorder category, instead placing them each into a separate classification system. As research and clinical experience have revealed distinct features and treatment approaches for both OCD and PTSD, the reclassification ensures more targeted and appropriate treatment options for individuals with these disorders.

The following table shows the difference in anxiety disorder classification between the previous version, the *DSM-IV* (with text revisions), and the current *DSM-5*.

DSM-IV-TR (2000)	DSM-5 (2013)
Anxiety disorders	**Anxiety disorders**
Panic disorder	Separation anxiety disorder
Agoraphobia	Selective mutism
Selective mutism	Specific phobia
Specific phobia (simple phobia)	Social anxiety disorder (social phobia)
Social phobia (social anxiety disorder)	Panic disorder
Obsessive-compulsive disorder	Panic attack (specifier)
Posttraumatic stress disorder	Agoraphobia
Acute stress disorder	Generalized anxiety disorder
Generalized anxiety disorder	**Obsessive-compulsive and related disorders**
	Trauma and stressor-related disorders

While these diagnoses may have changed in classification over the years, anxiety is still a common factor in all of them. This may be why anxiety disorders are the world's most prevalent mental disorders, affecting over 40 million adults in the United States (National Institute of Mental Health, 2023). In the following section, we will discuss some of the most common types of anxiety.

Generalized Anxiety Disorder

The most common anxiety disorder in older adults is GAD, which is characterized by persistent and excessive worry about a number of different things. Clients with GAD may anticipate disaster or be overly concerned about money, health, family, work, or other issues. GAD produces chronic, exaggerated worrying about everyday life. Sometimes just the thought of getting through the day produces anxiety (Noyes, 2001).

Clients with GAD find it difficult to control their worry, and this worry can consume hours each day, making it hard to concentrate or finish daily tasks. They don't know how to stop the worry cycle and feel it is beyond their control, even though they usually realize that their anxiety is more intense than the situation warrants. Some believe that worry prevents bad things from happening, so they view it as risky to give up worry (Noyes, 2001). However, their anxiety can produce a number of physical symptoms such as headaches, dizziness, tension, or nausea. The good news is that they found you as a therapist, and you are using this workbook to assist them in creating a healing toolkit that will help them to function socially, have a full and meaningful life, and face their fears!

> ― CASE STUDY ―
> ## Generalized Anxiety Disorder
>
> George is a 59-year-old man who reports that his biggest problem is constant worrying. He worries all of the time and states, "You name it, I worry about it." For example, he tells you he feels the same amount of worry about his wife's cancer diagnosis eight months ago and the pile of laundry on the floor that needs to be cleaned. He recognizes that his wife is much more important than a pile of laundry, but he is bothered that both cause him similar levels of worry. He reports having excessive and uncontrollable worry most days, all day long. He also describes difficulty falling asleep, ruminative what-if thoughts, irritability, trouble focusing at work, and significant back and muscle tension. George has struggled with anxiety for as long as he can remember, noting that his mother used to call him a "worry wart." His worrying has worsened since his wife's diagnosis.
>
> **Possible diagnosis:** Generalized anxiety disorder
>
> **Anxiety symptoms:**
> - Ruminative thoughts
> - Irritability
> - Sleep difficulties
> - Excessive and constant worry
> - Lack of concentration
>
> **Treatment recommendations:** George appears to be experiencing an exacerbation of anxiety symptoms due to the stress of his wife's diagnosis. He would benefit from therapy once per week, specifically person-centered CBT with a focus on mindfulness-based tools and techniques to help calm the mind and body. Assist George in challenging his automatic negative thoughts and practicing tools to self-soothe and regulate his nervous system.

Before moving forward with these treatment recommendations, it's important to assess your client's level of anxiety, as not everyone experiences anxiety the same way. You can use the following screening tool by Spitzer and colleagues (2006) to assess the severity of your client's level of anxiety. If your client has a dual diagnosis with PTSD, make sure to also screen for symptoms of hyperarousal or hypoarousal using the *Window of Tolerance Assessment* in chapter 4.

THERAPEUTIC WORKSHEET

GAD-7 Anxiety*

Over the last two weeks, how often have you been bothered by the following problems?	Not at all	Several days	More than half the days	Nearly every day
1. Feeling nervous, anxious, or on edge	0	1	2	3
2. Not being able to stop or control worrying	0	1	2	3
3. Worrying too much about different things	0	1	2	3
4. Trouble relaxing	0	1	2	3
5. Being so restless that it is hard to sit still	0	1	2	3
6. Becoming easily annoyed or irritable	0	1	2	3
7. Feeling afraid, as if something awful might happen	0	1	2	3

Column totals _____ + _____ + _____ + _____

= *Total score*** _____

If you checked off any problems, how difficult have these made it for you to do your work, take care of things at home, or get along with other people?

☐ Not difficult at all   ☐ Somewhat difficult   ☐ Very difficult   ☐ Extremely difficult

---

* Spitzer, R. L., Kroenke, K., Williams, J. B. W., & Löwe, B. (2006). A brief measure for assessing generalized anxiety disorder. *Archives of Internal Medicine, 166*(10), 1092–1097. https://doi.org/10.1001/archinte.166.10.1092

** Providers can access the scoring key for this assessment at https://adaa.org/sites/default/files/GAD-7_Anxiety-updated_0.pdf

High-Functioning Anxiety

Another common type of anxiety that is actually not recognized in the *DSM-5* is known as high-functioning anxiety. High-functioning anxiety is a lesser-known form of anxiety that characterizes individuals who appear outwardly successful and accomplished but struggle internally with persistent feelings of anxiety, self-doubt, and the fear of not measuring up. They may feel extremely uncomfortable inside and struggle with significant self-criticism, even if they appear to "have it all together." Many people who show symptoms of high-functioning anxiety are diagnosed with GAD and can usually maintain a high level of functionality in various aspects of their lives (Hubbard, 2023).

Individuals with high-functioning anxiety often seem to excel in multiple areas of their life. They may hold high-ranking positions, receive accolades for their work, or be seen as role models by their peers, but this seemingly successful facade often masks the internal struggle that they face on a daily basis. While they may appear to be doing well, they are actually constantly worrying about their performance, seeking validation from others, and fearing failure. Because they hold themselves to high standards and set unrealistic expectations, they may push themselves to work harder, take on more responsibilities, and constantly strive for a level of perfection that is unattainable. While this drive and ambition can be seen as admirable, it can also take a toll on their mental health (Shafir, 2022).

—CASE STUDY—

High-Functioning Anxiety

Sarah is a 40-year-old single woman who reports that she's been feeling anxious and lonely off and on since her birthday a couple months ago. She has always been vibrant, adventurous, and career driven. However, as she approaches midlife, she finds herself dealing with new fears and concerns. She has developed a deep-rooted fear of aging and has been feeling pressure to look younger. She recently decided to get a consultation for a facelift and spends an excessive amount of time at the gym.

Sarah also reports a constant worry about her well-being and an overwhelming belief that she is constantly on the verge of a catastrophic health event. She constantly feels like she is having chest pains and believes she is having a heart attack, even though she has seen multiple cardiologists and all tests have come back clear. Additionally, she has been grappling with the fear of being alone forever, stating, "I'm not married, and I have no kids. Why doesn't anyone want me? I'm such a loser." Sarah immerses herself in work to distract herself from this loneliness. She has continuously received accolades from her boss for all of the long hours she's put in, yet she still feels inadequate. She even received a very high-ranked promotion after only a year at her job, which is the only thing in her life she is proud of. Sarah states that she just wants to have "the perfect life" but that she's too old to find love.

While most of her days are consumed with work duties, Sarah still tries to prioritize time for social fun. Yoga and hiking are two of her favorite activities that she knows calm her nervous system, but even during these activities her mind races with intrusive thoughts about death: "Am I dying?" "What if someone close to me dies?" "What if I have a heart attack right now?" Her friendships create a sense of calm for her, but sometimes being around her married friends with kids make her feel inadequate and sad. Some days Sarah feels a sense of impending doom, like she's on the verge of losing control, but other days are filled with hope and happiness. Sarah is looking for a safe space to process her feelings, quiet her anxious thoughts, and reflect on her decisions in life.

Possible diagnosis: High-functioning anxiety, r/o illness anxiety disorder, generalized anxiety disorder

Anxiety symptoms:

- Chest pains
- Constant worry about her well-being
- Catastrophic thinking and fears of the future
- Perfectionism
- Pressure to look younger
- Feelings of inadequacy
- Overworking to distract herself from loneliness
- Ruminative thoughts about her health
- Anxious thoughts about the aging process

Treatment recommendations: Sarah would benefit from a consistent mindfulness practice and cognitive restructuring to combat catastrophic thinking and fears of the future. Somatic exercises—such as body scans, breathwork, and yoga (one of her preferred activities)—will help increase emotional and bodily awareness. Future visualization meditations will help her focus on more calming and positive images about her future and can enhance her relaxation skills. Self-love and affirmation work should also be considered to help reprogram her brain to focus less on the inner critic and more on compassionate and loving self-talk.

The following tools can help you identify symptoms of high-functioning anxiety and related perfectionistic tendencies in your clients.

THERAPEUTIC WORKSHEET

High-Functioning Anxiety

The following list includes some of the most common characteristics of individuals with high-functioning anxiety. Place a check mark next to any of the characteristics you identify with.

- ☐ Have a hard time saying no
- ☐ Often seek reassurance from others
- ☐ Struggle with feelings of self-doubt
- ☐ Fear letting others down
- ☐ Meticulously double-check minor details
- ☐ Constantly worry about job performance
- ☐ Hide your anxiety from others
- ☐ Appear calm and confident without feeling it on the inside
- ☐ Have difficulty relaxing
- ☐ Felt insecurity as a child
- ☐ Grew up with caregivers who were anxious or had high expectations of you
- ☐ Seek validation from others
- ☐ Fear failure
- ☐ Are a high achiever
- ☐ Constantly strive for perfection
- ☐ Are highly successful in your career
- ☐ Hold yourself to high standards
- ☐ Fear looking inadequate or foolish to others
- ☐ Have frequent racing thoughts
- ☐ Have difficulty sleeping
- ☐ Constantly overthink things

PSYCHOEDUCATIONAL WORKSHEET

Perfectionism Looks Like . . .

The following list includes some of the most common characteristics of perfectionism. Place a check mark by any characteristics that you identify with.

- ☐ Have high standards and expectations
- ☐ Feel pressure to live up to high expectations
- ☐ Need clear organization and structure
- ☐ Have an exaggerated fear of failure
- ☐ Are ambitious and driven
- ☐ Experience high levels of self-doubt and insecurity
- ☐ Have difficulty overlooking small mistakes
- ☐ View any mistake as failure or incompetence
- ☐ Fear you'll be rejected or judged because of mistakes
- ☐ Spend excessive time, effort, or energy to improve or reduce mistakes
- ☐ Excessively ruminate or engage in self-criticism
- ☐ Are always over-preparing or creating a plan
- ☐ Need exact rules, expectations, and instructions
- ☐ Are hypersensitive to criticism and negative feedback
- ☐ Have rigid black-or-white thinking patterns
- ☐ Have self-worth or self-esteem that is contingent upon success

CHAPTER 4

Understanding the Trauma Response

Another important consideration to the success of anxiety healing is the presence of trauma, which can significantly contribute to the development and maintenance of anxiety disorders. Many times, people hear the word *trauma* and automatically think of big "T" traumas or catastrophic events, such as pandemics, terrorist attacks, natural disasters, sexual abuse, or the death of a child. While these are very traumatic experiences, it's important to acknowledge the existence of little "t" traumas that happen in everyday life as well. Little "t" traumas are distressing events that are comparatively smaller in scale but still disrupt an individual's ability to cope. This can include relationship conflict, legal difficulties, financial concerns, bullying, and more.

It is crucial for therapists to understand that no two traumas are the same and that trauma can occur as a result of a one-time event or can stem from ongoing adverse experiences. For example, you might have one client who experienced a devastating earthquake that destroyed their home and caused the loss of loved ones, leaving them with lasting emotional and psychological scars. You might have another client who endured years of physical and emotional abuse as a child at the hands of their caregiver, leading to difficulty in forming healthy relationships, low self-esteem, and consistent nightmares of the abuse. Despite the disparate nature of their trauma experiences, both of these clients may meet criteria for GAD and PTSD as a result.

Remember that regardless of the type of trauma someone has endured, each person's experience is unique and should be treated as such. As therapists, it is important to approach each case with an open mind and to recognize that all trauma is of equal importance. This means providing a safe, supportive, and nonjudgmental environment where healing can take place.

Anxiety versus Trauma

Anxiety and trauma are closely related and are often confused, but they are distinct concepts. It is important to understand the difference in order to effectively address and heal symptoms that your client may be experiencing. Anxiety is a response to a perceived threat, whether real or imagined. As discussed in chapter 2, when someone's body senses danger, the sympathetic nervous system activates the fight-or-flight response to help them confront or flee the perceived threat at hand. The physical symptoms that accompany this fight or flight are characteristic of anxiety.

Unlike anxiety, trauma is not a response to a perceived threat but the result of actual danger or a harmful event that has occurred in someone's life. While anxiety can be triggered by everyday stresses and worries, trauma is rooted in past experiences that have had a profound impact on the psyche. These experiences can leave clients feeling helpless, overwhelmed, and disconnected from themselves and others. The effects of trauma can be long-lasting and can greatly affect an individual's ability to function and enjoy life.

As a clinician, it's important to explore the profound impact that trauma can have on someone's perception of danger. When a client experiences a traumatic event, their brain starts to associate everyday situations with a threat to their life. This creates a constant state of anxiety and hypervigilance, making it difficult for them to feel safe in their own skin. This is especially the case for clients who have experienced childhood trauma, as children rely on their caregivers to provide a sense of safety. When this sense of safety is compromised, their ability to self-regulate and tolerate stress is compromised. Imagine you have a client who grew up in a household where their caregivers were constantly fighting, neglectful, or abusive. Perhaps another client has survived living in an environment where violence was an everyday occurrence. These clients' nervous systems learn to be on high alert constantly, leaving them unable to trust others or tolerate life's inevitable stressors. In adulthood, the effects of these traumatic experiences can persist—manifesting as a variety of anxiety disorders and other mental health issues.

Through more specialized trauma therapies, such as trauma-focused CBT (TF-CBT), eye movement desensitization and reprocessing (EMDR), cognitive processing therapy, and prolonged exposure therapy, clients with traumatic experiences can learn to rewire their brains and develop healthy self-regulation skills. These therapies specifically target the traumatic experiences and work toward reducing distressing symptoms and promoting healing. If you are working with a client who has experienced trauma or is diagnosed with PTSD, this workbook should only be used as a supplemental form of healing, but it can be successful treatment as long as your client remains within their window of tolerance, which will be discussed next.

Window of Tolerance

If you are working with a client who fits the criteria for PTSD, it is important to carefully evaluate their "window of tolerance" when using this workbook for treatment. The window of tolerance is a concept originally developed by Dr. Dan Siegel (1999); it describes the optimal zone of arousal in which a person can manage the stressors of everyday life. When an individual is operating within this zone—or "inside" the window of tolerance—their thoughts are clear and rational, and their emotions are regulated and controlled. If a client is in this optimal arousal zone, they will:

- Feel present
- Feel safe
- Be resilient
- Be able to adapt to fit the situation
- Experience empathy
- Feel and think simultaneously
- Effectively communicate with others
- Feel open and curious (versus judgmental and defensive)
- Be able to regulate their emotions
- Have an awareness of boundaries (their own and others')
- Remember to use their healing toolkit more easily

However, when a client tips outside of their window of tolerance, the prefrontal cortex shuts down, affecting their ability to think rationally and manage difficult emotions. This can cause arousal levels to rapidly increase (*hyperarousal*) or decrease (*hypoarousal*), leading the individual to feel out of control and become "stuck" outside of their window for quite some time. When this occurs, the client can experience difficulty sleeping, lack of concentration, and an inability to manage emotions. Physically, their body may be tense and on the brink of an explosion, which can result in angry outbursts and hostility. Let us look at hyperarousal and hypoarousal in more detail.

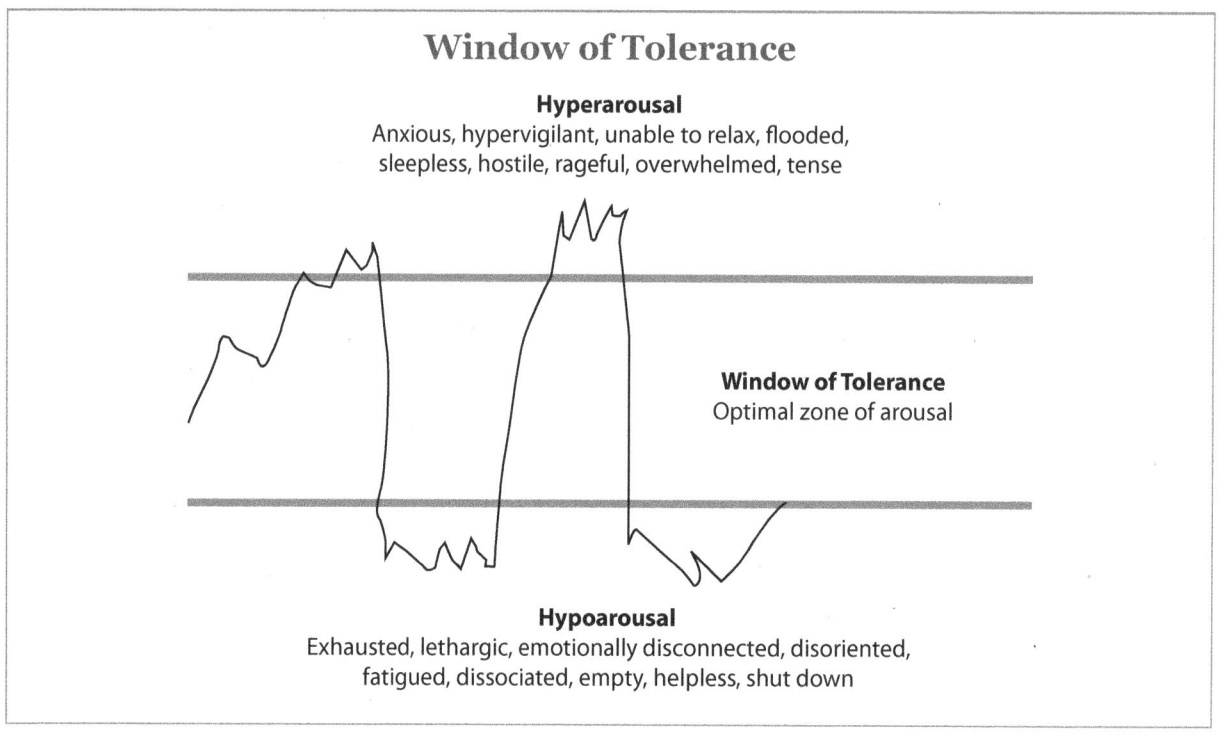

Understanding Hyperarousal

When someone is *above* their window of tolerance, they are in a state of hyperarousal, which can manifest in physical symptoms such as rapid heart rate, shallow breathing, tense muscles, and sweating. Mentally, hyperarousal may lead to difficulty concentrating, racing thoughts, and irritability. It can occur quickly and can be hard to prevent or recover from. Some common symptoms that may occur in this state include:

- Angry outbursts or agitation
- Muscle tension or shaking
- Anxiety, fear, or panic
- An inability to rest
- Emotional overwhelm
- Racing thoughts
- Sleep issues
- Defensiveness
- Hypervigilance
- Intrusive images
- Difficulty being in busy or crowded environments
- Difficulty concentrating
- Chronic pain
- An elevated heart rate
- Hypertension
- Impulsivity
- Restlessness

Clients can get sent into a state of hyperarousal when they are stressed or encounter reminders of traumatic memories. It's important to recognize that hyperarousal is a natural response to perceived threats or overwhelming emotions. The body goes into "alarm" mode in an attempt to respond to the perceived danger and protect itself (Gill, 2017). For clients who have experienced trauma, though, this alarm mode becomes overactive—leading them to react to triggers that do not represent a true threat to their safety.

—CASE STUDY—
Hyperarousal

Joan is a 45-year-old woman who recently had a panic attack and was recommended to seek therapy by her doctor. She has always struggled with the holiday season ever since she was nine years old, when her father had a heart attack and died in front of her on Christmas Eve. Since then, Joan has tried her best to distract herself and keep herself busy during this time of year. However, as a people pleaser, Joan has a difficult time saying no anytime she is invited somewhere or asked to help someone. As this year's holiday season approaches, she finds herself not only more restless than past years but also constantly worried that she is having heart problems.

Joan reports feeling restless most days, as she is juggling multiple holiday parties, planning a vacation, hosting playdates for her children at her house, volunteering at her children's school, shopping for gifts, and taking care of her elderly mother, who lives with the family. Joan is also barely sleeping due to nightmares about the day her father died, along with constant rumination about her own death. She has frequent anger outbursts, struggles with racing thoughts, and is very irritable. She also experiences physical symptoms such as dizziness, crying spells, and headaches.

Last month, Joan was having severe chest pains and became convinced that she was dying of a heart attack. Of note, Joan is almost the same age as her father was when he had his heart attack. After a visit to the emergency room and multiple tests, the doctor reported that her heart looked very healthy and that the symptoms were indicative of a panic attack. Although Joan is seeking therapy for her recent panic attack, she is still convinced the doctors missed something.

Possible diagnosis: PTSD, generalized anxiety disorder, illness anxiety disorder

Hyperarousal symptoms:

- Frequent anger outbursts
- Racing thoughts
- Irritability
- Dizziness
- Crying spells
- Headaches
- Intrusive images of her father's death
- Recent panic attack

Treatment recommendations: Joan appears to be experiencing anxiety symptoms as she approaches the age her father was when he experienced a heart attack. It appears that her father's death may have had a bigger impact on her than she may have realized, especially as an adult who is becoming aware of her own mortality. She may require multiple sessions of specialized trauma treatment (e.g., TF-CBT, EMDR) per week.

Understanding Hypoarousal

When someone is *below* their window of tolerance, they are in a state of hypoarousal, which can manifest as emotional numbness, physical lethargy, and a lack of desire to carry out tasks they normally enjoy. In this state, it's important to gently challenge clients to engage in activities that bring them joy and connection without pushing themselves too far beyond their window of tolerance (Gill, 2017). Common symptoms that someone may experience in this state include:

- Depression
- Low energy
- Memory loss
- Slow cognitive processing ("I can't think")
- Feelings of shame
- Emptiness
- Feelings of disconnection
- Feelings of being shut down
- Helplessness
- Inability (or lack of desire) to speak
- Blank stare
- Dissociation
- Slow digestion
- Low blood pressure

Like hyperarousal, clients can often be triggered into a state of hyperarousal when they feel threatened, recount traumatic memories, or experience emotions associated with a past trauma. Even a perceived threat can be enough to send someone into a state of shutdown or even dissociation. You might recall from chapter 2 that the instinct to freeze, shut down, and become emotionally withdrawn is an adaptive response in the face of a true life-threatening situation, but when it persists in the absence of danger, it can contribute to chronic anxiety.

—CASE STUDY—
Hypoarousal

Mark is a 35-year-old man who is seeking therapy at the request of his wife after he recently stated to her, "I don't know why I'm even here on this earth." He reports feeling lethargic and apathetic about life lately. Along with constant fighting in his relationship, he has been struggling with the loss of his job last month. Mark was let go because he consistently missed important meetings and began forgetting to complete important work tasks. Although he has always been a top performer at his job, loves what he does, and goes above and beyond in his professional and personal life, he has recently been feeling disconnected, empty, and numb. He states that he has been trying to avoid having to talk about all of this, hoping it would all just go away.

Mark also reports feeling worthless and like nothing he ever does is good enough, especially after running into his father a couple of months ago. Prior to this incident, Mark hadn't seen his father since he left his childhood home over 17 years ago, following what he describes as "some of the darkest days of my life." Mark reports being physically abused by his father for most of his childhood and always being told that he was "nothing." Since seeing his father, Mark has had recurring nightmares about the abuse that keep him from falling asleep. He continues to isolate himself, has no appetite, and hardly gets out of bed.

Possible diagnosis: PTSD, major depressive disorder

Hypoarousal symptoms:

- Emotional numbness
- Memory loss
- Feeling of being shut down
- Slowed cognitive processing
- Feelings of emptiness and disconnection

Treatment recommendations: Mark appears to be having a PTSD response related to his recent interaction with his abusive father. It seems that he may have dissociated from this trauma and buried any associated emotions for quite some time. He may require multiple sessions of specialized trauma treatment (e.g., TF-CBT, EMDR) per week.

THERAPEUTIC WORKSHEET

Window of Tolerance Assessment

Use this worksheet to identify and recognize the symptoms of hyperarousal and hypoarousal. Becoming more self-aware of these symptoms will help you assess if you are currently above (hyperarousal), below (hypoarousal), or within your window of tolerance.

In the following table, circle the symptoms you are feeling, and use this information to rate your level of hyperarousal or hypoarousal.

		Symptoms	
1 **Hyperarousal** *(more anxious)*	**Hyperarousal** • Increased arousal • Anxiety, anger, or loss of control • Impulse to fight or run away	• Anxiety • Impulsivity • Intense reactions • Lack of emotional safety • Hypervigilance • Intrusive thoughts • Fear • Shaking	• Rigidity • Defensiveness • Anger or rage • Physical and emotional aggression • Obsessive thoughts and behaviors • Emotional outbursts • Racing thoughts
5 **Window of Tolerance** *(calm and regulated)*	**Window of Tolerance** • Balance and calm • Stability and control • Ability to function effectively and handle the adversities that life throws at you	When you're in your window of tolerance, you feel like you can deal with whatever is happening in your life. You might feel stress or pressure, but it doesn't bother you too much. This is the ideal place to be. • Emotional regulation • Ability to self-soothe • Effective functioning	
10 **Hypoarousal** *(more depressed)*	**Hypoarousal** • Decreased arousal • Depression, emotional numbness, or exhaustion • Shutdowns or freezing	• Feeling of being on autopilot • Lack of energy • Inability to think or respond • Memory loss • Numbness • Feeling of zoning out	• Shutdowns • Reduced physical movement • Shame • Depression • Difficulty engaging in coping skills • Low levels of energy

My rating is: _____

What symptoms are the most severe? _____

_____

Trauma and the Window of Tolerance

Trauma can narrow a client's window of tolerance, making it difficult for them to regulate their emotions and stay within the "optimal zone" where they can function effectively. Essentially, certain emotions and situations feel much more intense and difficult to manage. Even seemingly minor stressors can cause clients to dissociate, get angry, or feel anxious, leading to states of hyperarousal or hypoarousal. This makes it harder for clients who are outside their window of tolerance to make therapeutic progress.

Therefore, it is vital to assess your clients in terms of their window of tolerance before moving forward with clinical treatment. Clients who are within their window of tolerance are much more likely to progress in their healing than those who are stuck in a state of hyperarousal or hypoarousal. You can use the following *PTSD Checklist for* DSM-5 (PCL-5; Weathers et al., 2013) to evaluate for the presence of PTSD among clients with a history of trauma. A cutoff score between 31–33 is indicative of a probable diagnosis. If you find that a client would benefit from a higher level of care, make sure to refer them to the appropriate trauma specialist.

THERAPEUTIC WORKSHEET

PTSD Checklist for *DSM-5*

This questionnaire asks about problems you may have had after a very stressful experience involving actual or threatened death, serious injury, or sexual violence. It could be something that happened to you directly, something you witnessed, or something you learned happened to a close family member or close friend. Some examples are a serious accident; fire; disaster such as a hurricane, tornado, or earthquake; physical or sexual attack or abuse; war; homicide; or suicide.

First, please answer a few questions about your worst event, which for this questionnaire means the event that currently bothers you the most. This could be one of the examples above or some other very stressful experience. Also, it could be a single event (for example, a car crash) or multiple similar events (for example, multiple stressful events in a war zone or repeated sexual abuse).

Briefly identify the worst event (if you feel comfortable doing so):

_____
_____
_____
_____
_____
_____

How long ago did it happen? _____ (please estimate if you are not sure)

Did it involve actual or threatened death, serious injury, or sexual violence?

☐ Yes ☐ No

How did you experience it?

☐ It happened to me directly.
☐ I witnessed it.
☐ I learned about it happening to a close family member or close friend.
☐ I was repeatedly exposed to details about it as part of my job (for example, paramedic, police, military, or other first responder).
☐ Other (please describe): _____

If the event involved the death of a close family member or close friend, was it due to some kind of accident or violence, or was it due to natural causes?

- ☐ Accident or violence
- ☐ Natural causes
- ☐ Not applicable (the event did not involve the death of a close family member or close friend)

Second, below is a list of problems that people sometimes have in response to a very stressful experience. Keeping your worst event in mind, please read each problem carefully and then circle one of the numbers to the right to indicate how much you have been bothered by that problem *in the past month*.*

In the past month, how much were you bothered by:	Not at all	A little bit	Moderately	Quite a bit	Extremely
1. Repeated, disturbing, and unwanted memories of the stressful experience?	0	1	2	3	4
2. Repeated, disturbing dreams of the stressful experience?	0	1	2	3	4
3. Suddenly feeling or acting as if the stressful experience were actually happening again (as if you were actually back there reliving it)?	0	1	2	3	4
4. Feeling very upset when something reminded you of the stressful experience?	0	1	2	3	4
5. Having strong physical reactions when something reminded you of the stressful experience (for example, heart pounding, trouble breathing, sweating)?	0	1	2	3	4
6. Avoiding memories, thoughts, or feelings related to the stressful experience?	0	1	2	3	4
7. Avoiding external reminders of the stressful experience (for example, people, places, conversations, activities, objects, or situations)?	0	1	2	3	4

---

* Providers can access the scoring key for this assessment at https://www.ptsd.va.gov/professional/assessment/documents/using-PCL5.pdf

In the past month, how much were you bothered by:	Not at all	A little bit	Moderately	Quite a bit	Extremely
8. Trouble remembering important parts of the stressful experience?	0	1	2	3	4
9. Having strong negative beliefs about yourself, other people, or the world (for example, having thoughts such as: I am bad, there is something seriously wrong with me, no one can be trusted, the world is completely dangerous)?	0	1	2	3	4
10. Blaming yourself or someone else for the stressful experience or what happened after it?	0	1	2	3	4
11. Having strong negative feelings such as fear, horror, anger, guilt, or shame?	0	1	2	3	4
12. Loss of interest in activities that you used to enjoy?	0	1	2	3	4
13. Feeling distant or cut off from other people?	0	1	2	3	4
14. Trouble experiencing positive feelings (for example, being unable to feel happiness or have loving feelings for people close to you)?	0	1	2	3	4
15. Irritable behavior, angry outbursts, or acting aggressively?	0	1	2	3	4
16. Taking too many risks or doing things that could cause you harm?	0	1	2	3	4
17. Being "superalert" or watchful or on guard?	0	1	2	3	4
18. Feeling jumpy or easily startled?	0	1	2	3	4
19. Having difficulty concentrating?	0	1	2	3	4
20. Trouble falling or staying asleep?	0	1	2	3	4

PART III

Untangling Anxiety

CHAPTER 5

The Anxiety Cycle

Anxiety may manifest differently for everyone, but there is typically a universal four-stage cycle that people go through. The cycle starts with an event or situation that creates such an intense fear or worry that is so destabilizing (stage 1) for an individual that they have to avoid the situation (stage 2) in order to escape these extremely uncomfortable emotions. This avoidance provides some short-term relief from the intense discomfort (stage 3), but since this relief is temporary, the anxiety comes back (stage 4) worse than before. As a result, the individual feels out of control and their fears grow increasingly powerful, making the avoidance difficult to resist. Let's take a closer look at all four stages and examine what can help break this cycle.

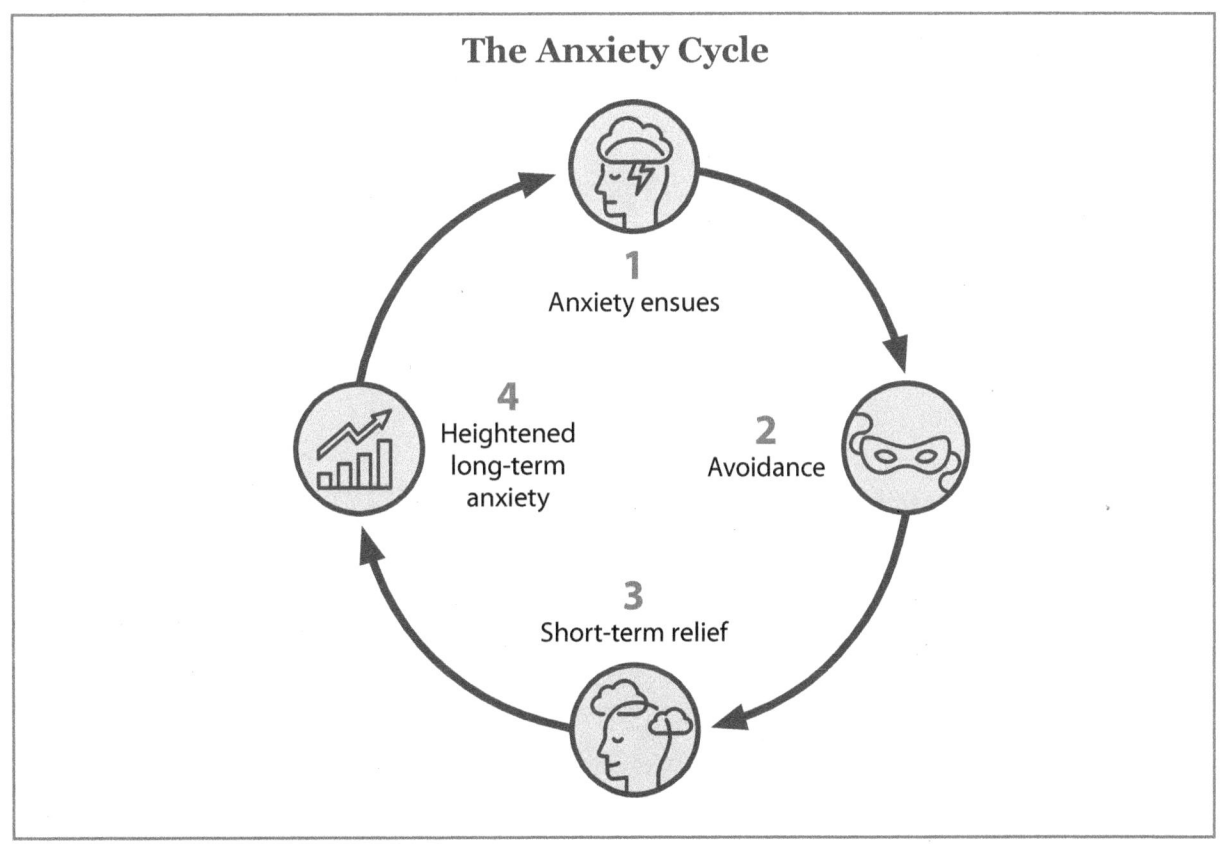

Stage 1: Anxiety Ensues

The cycle begins when a client becomes triggered by a stressful event or situation, leading them to go into fight-or-flight mode. This precipitating event can be anything that the client finds triggering. For example, the client's anxiety may arise in response to an intrusive thought or image, a memory, certain physical sensations, a social situation, a relationship interaction, or even a post on social media. In turn, they may experience symptoms such as:

- Feeling nervous, restless, or on edge
- Feeling impending danger or doom
- An increased heart rate, fast breathing, sweating, or trembling
- Feeling weak and tired
- Trouble sleeping
- Excessive worry
- Gastrointestinal issues

The client's attention then focuses only on this provoking stimulus, which raises their stress level even more and contributes to more anxiety symptoms. This emotional response becomes too intense, which leads to the second stage of the cycle: avoidance (Nawal, 2023).

THERAPEUTIC WORKSHEET

The Anxiety Response

In the following table, circle the physical, emotional, and psychological symptoms you experience when you're anxious. These symptoms are indications that you may be in the beginnings of the anxiety cycle.

Physical	Emotional	Psychological
Migraines	Sad	Insomnia
Sweating	Angry	Hyposomnia
Upset stomach	Tired	Gambling
Diarrhea	Worried	Increased substance use
Increased heart rate	Irritable	Impulsivity
Muscle tension	Scared	Inability to concentrate
Body weakness	Frustrated	Isolation
Low immune system	Lonely	Nail biting
Low/high sex drive	Excited	Hair pulling, skin picking
Fast breathing	Disappointed	Procrastination
Low energy	Jealous	Forgetfulness
Weight loss or gain	Embarrassed	Focusing on the negative
Other: _____	Other: _____	Other: _____

Stage 2: Avoidance

When anxiety symptoms become too uncomfortable, clients will do anything to avoid the activating event that caused anxiety in the first place. Avoidance may take the form of actions like:

- Skipping class to avoid giving a presentation
- Using drugs or alcohol to numb feelings
- Procrastinating doing challenging tasks
- Avoiding confrontation in close relationships
- Canceling plans or ignoring phone calls/texts
- Leaving early or removing yourself from a social situation
- Sleeping or napping to escape the situation

Avoidance may also take the form of safety behaviors, which are subtle techniques that people use to prevent, escape, or cope with anxiety-provoking situations. These may include carrying around a comfort object (e.g., anti-anxiety medication, cell phone) at all times, googling information about feared symptoms, seeking reassurance from others, making sure a trusted loved one is with them when out in public, or always having an escape plan. In order to break this cycle, clients must face their fears by gradually confronting the anxiety-provoking situation in a slow and systemic way (Nawal, 2023).

THERAPEUTIC WORKSHEET

Safety Behaviors

One of the most harmful responses to anxiety is avoidance. Safety behaviors are any subtle actions used to avoid anxiety in certain situations. For example, someone who is anxious about socializing at a party might focus on their phone to discourage others from approaching. Although safety behaviors provide relief in the short term, they make anxiety worse in the long run. Place a check mark by any safety behaviors that you engage in.

- ☐ Participating in superstitious behaviors before, during, or after situations that increase anxiety
- ☐ Avoiding places or situations that tend to increase anxiety
- ☐ Carrying around a "security object" in situations that tend to increase anxiety
- ☐ Frequently visiting the doctor due to any uncomfortable physical sensations
- ☐ Frequently checking your vitals (e.g., heart rate, blood pressure, oxygen levels)
- ☐ Checking if the door is locked multiple times throughout the night and staying awake for long periods of time to make sure you're "safe" at home
- ☐ Rewriting and rereading texts or emails before sending them
- ☐ Always calling another person during a situation that elicits anxiety
- ☐ Being overly prepared for school, work, presentations, or meetings
- ☐ Constantly rehearsing conversations in your mind before they happen
- ☐ Taking a bottle of anti-anxiety medication everywhere you go outside of the home
- ☐ Leaving the house only if you have a certain object with you or if someone else is with you
- ☐ Having an escape plan anytime you go somewhere that elicits anxiety
- ☐ Frequently checking your phone during social interactions
- ☐ Drinking alcohol or using recreational drugs or other substances to curb anxiety
- ☐ Other: _____

BETWEEN-SESSION WORKSHEET

Daily Check-In: Safety Behaviors

Use this daily check-in to record any safety behaviors you engage in when anxiety arises. If needed, you can use the checklist from the *Safety Behaviors* worksheet for some examples of common safety behaviors. If you're able to identify any triggers that increase your safety behaviors, write them in the appropriate box.

Day/Time	Safety Behaviors	Triggers
Monday		
Tuesday		
Wednesday		
Thursday		
Friday		
Saturday		
Sunday		

Stage 3: Short-Term Relief

While avoidance behaviors provide temporary relief, they may not always be helpful in the long run. This is because the more that someone avoids a situation, even if it feels better over the short run, the more they are reinforcing this behavior. According to The OCD and Anxiety Center, "Negative reinforcements can be just as powerful of motivators as positive reinforcements." In fact, we know that "any behavior that is rewarded, whether it is negatively or positively rewarded, is more likely to continue. People tend to be more familiar with positive reinforcements, such as rewards. Negative reinforcement means that an aversive stimulus is removed and as a result, relief is experienced. In this context, the aversive stimulus is whatever is provoking the anxiety" (Butterfield, 2021). In other words, even though negative reinforcement may decrease anxiety at the moment, avoidance becomes the only way to cope, and the relief becomes increasingly short-lived as untreated symptoms worsen. The bottom line: Avoidance does not help. It worsens symptoms and leaves clients feeling powerless against their anxiety.

Stage 4: Heightened Long-Term Anxiety

The cycle completes with continuous, long-term heightened anxiety. When an individual engages in avoidance behaviors to manage anxiety, their ability to tolerate distress decreases. The fear that initially led to avoidance worsens, and their brain learns that avoidance is the only way to make the anxious symptoms go away. As a result, the anxiety becomes worse the next time, and avoidance becomes more likely. This repetitive cycle is what keeps clients stuck. As challenging as it may be, guiding clients to safely confront and face their fears will be the path toward treatment success.

THERAPEUTIC WORKSHEET

The Four Stages of Anxiety

The following diagram illustrates the four stages of anxiety: (1) anxiety ensues, (2) avoidance, (3) short-term relief, and (4) heightened long-term anxiety.

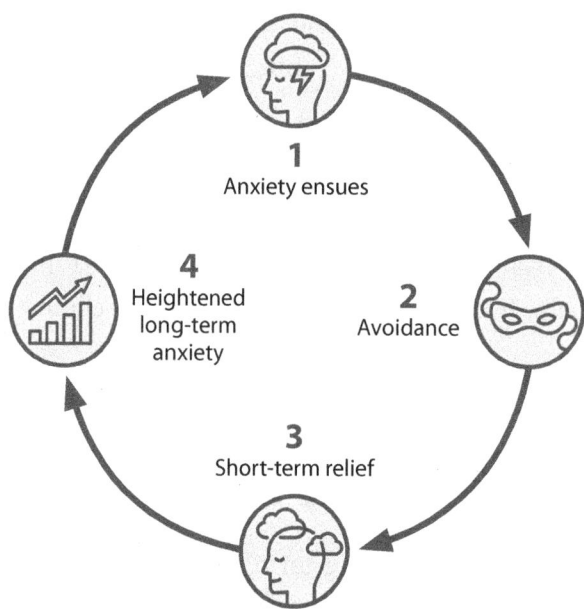

To learn more about your own cycle of anxiety, read the following example and then work through the four stages of anxiety based on your own experience.

Situation: Your friend wants to introduce you to a new group of people and invites you to a party.	
Stage 1: Anxiety ensues	**How you might feel:** The thought of socializing with new people makes your heart race. You can't stop thinking about the party, and you feel so nervous that you're going to say something stupid. You haven't been able to sleep because you've been thinking about it so much.
Stage 2: Avoidance	**What you might do:** To avoid the party, you say that an emergency has come up and you can't go.
Stage 3: Short-term relief	**How that makes you feel:** You feel relieved that you don't have to attend the party and meet new people.
Stage 4: Heightened long-term anxiety	**What might happen in the future:** Your friend reaches out again two weeks later to invite you to a movie night with the group from the party. The cycle of anxiety starts, and you feel the same symptoms of anxiety as you did before. To avoid movie night, you say that you are sick and can't join.

Now, let's practice with an example from your own life:

Situation:	
Stage 1: Anxiety ensues *What are your symptoms? How is it affecting your life?*	**How you might feel:**
Stage 2: Avoidance *What are your behaviors?*	**What you might do:**
Stage 3: Short-term relief *Do you feel relieved? How long does it last?*	**How that makes you feel:**
Stage 4: Heightened long-term anxiety *What might happen in the future? How intense is your anxiety when you try to confront the fear?*	**What might happen in the future:**

How to Manage the Cycle of Anxiety

Research has shown that the most successful treatment method for anxiety disorders is CBT, especially for clients who struggle with generalized anxiety (Borza, 2017; Mayo Clinic, 2018). This is because CBT is typically a short-term treatment that focuses on teaching the client specific skills to improve their symptoms. This work involves helping clients alter maladaptive emotional responses by changing irrational thought patterns and behaviors, then gradually confronting the activities they have avoided because of their anxiety. The next chapter dives more deeply into the CBT model and explains how to use it to treat anxious clients.

The Cognitive Behavioral Model

One of the most widely known treatments for anxiety disorders is CBT. The cognitive model was developed by Aaron Beck in the 1960s as a framework for understanding a person's mental distress. According to the model, thoughts influence emotions, which, in turn, affect the way people behave. CBT helps clients identify the core beliefs that underlie their irrational thought patterns so they can feel and behave differently. This action-oriented type of psychological treatment that explores the links between thoughts, emotions, and behaviors takes a more proactive approach toward changing disordered thinking and negative core beliefs (Cully & Teten, 2008).

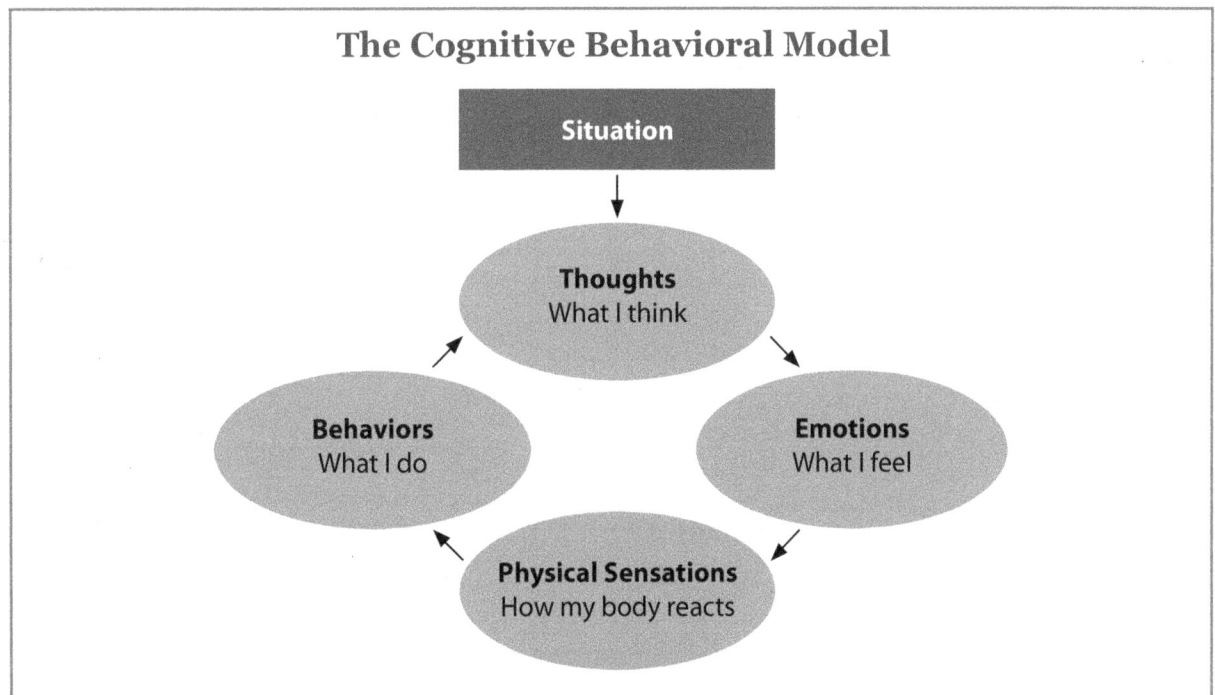

One of the key features of CBT is the establishment of a collaborative relationship between the client and you as the clinician. In this partnership, you work together as a team to identify maladaptive cognitions and behaviors, test their validity, and make necessary revisions. This collaborative approach is crucial in promoting effective healing and overcoming anxiety. Rather than being an authoritative figure, you act as a guide, working hand in hand with the client. This collaboration creates an atmosphere of trust and mutual respect, which helps the client feel comfortable opening up about their thoughts, emotions, and behaviors (Wright, 2006).

PSYCHOEDUCATIONAL WORKSHEET

The CBT Framework

The CBT model describes how your thoughts and emotions influence how you feel. It starts with a distressing situation or trigger → which causes you to have negative thoughts → which causes negative emotions and physical distress → which then leads to negative behaviors. Look at the example of the CBT model below and then complete your own model with an example from your own life. This will help you learn more about the thoughts, emotions, behaviors, and body sensations that you experience when anxious.

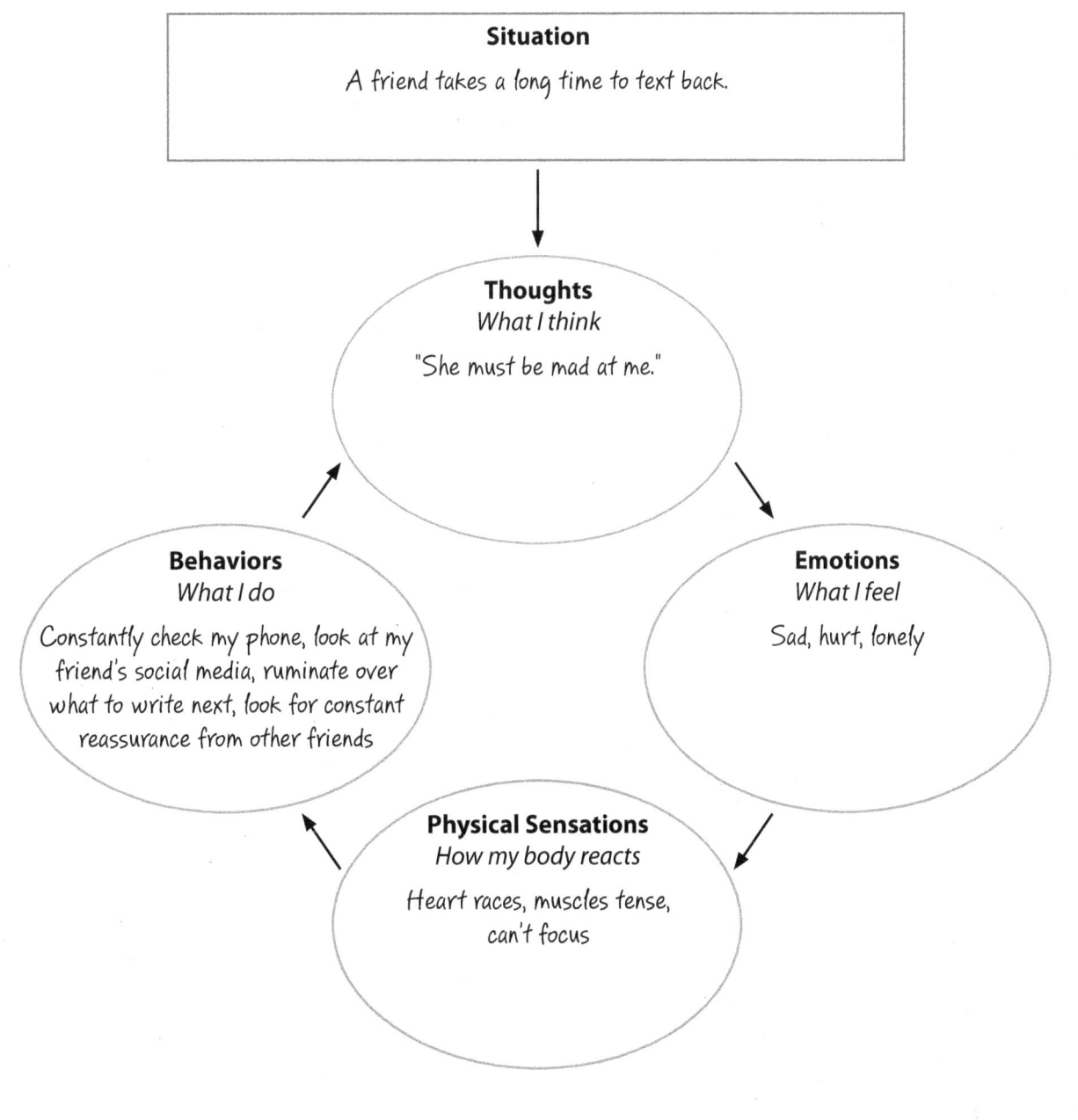

Copyright © 2024 Alison Seponara, *The Anxiety Healer's Guide for Clinicians*. All rights reserved.

Now practice with an example from your own life.

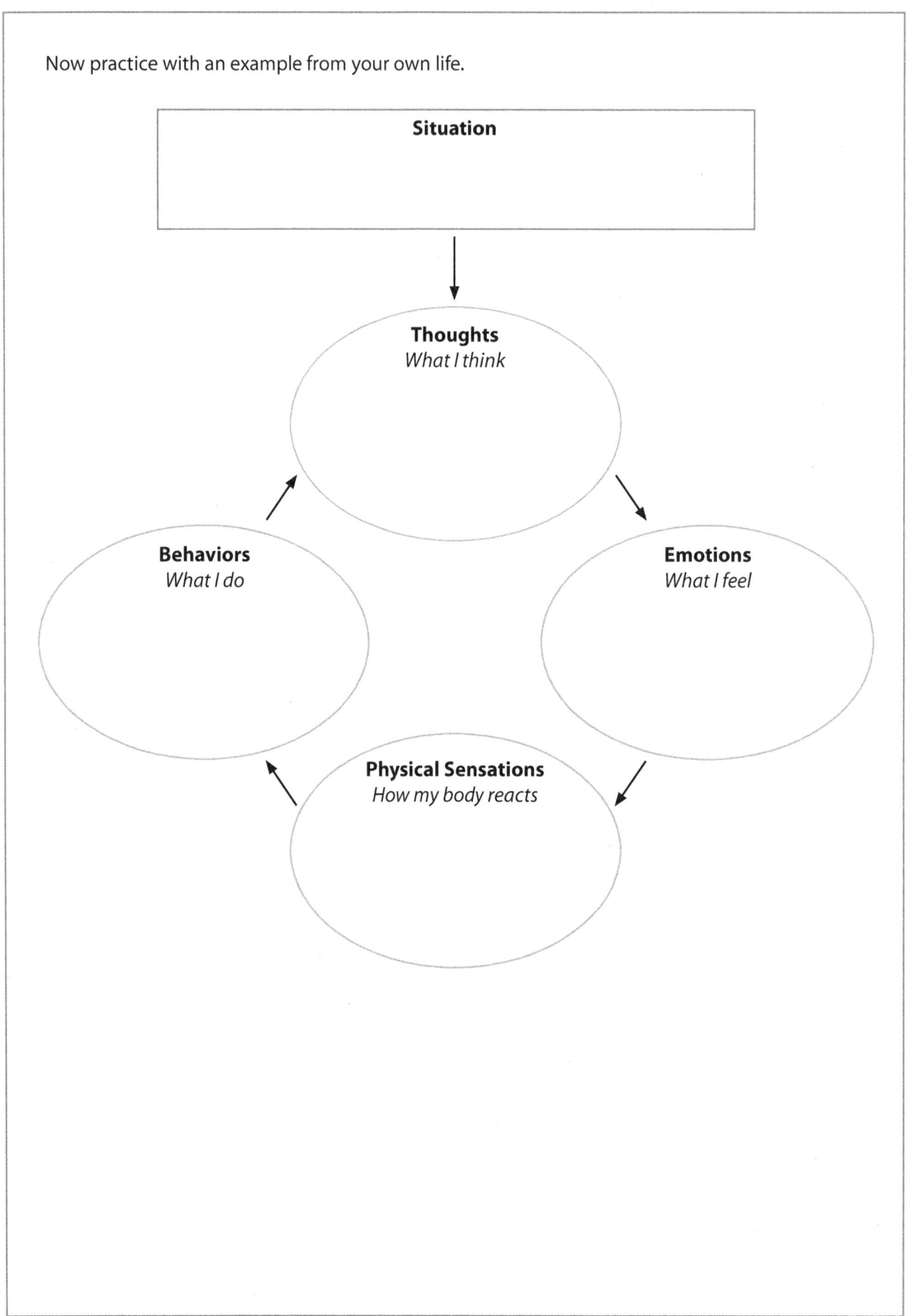

Levels of Cognition

As we've discussed, a core component of CBT involves helping clients alter their thoughts, or cognitions, so they can improve the way they feel and behave. When it comes to cognitions, Beck defined three major levels at which they can occur (Clark et al., 1999).

1. Full consciousness
2. Core beliefs
3. Automatic thoughts

Full consciousness is the state in which rational decisions are made with full awareness. When an individual is thinking rationally and is able to manage their emotions in a healthy way, they are thinking from a fully conscious mind. They are aware of the thoughts and beliefs that limit their ability to make conscious decisions and are able to face emotions head-on, without distorted thinking getting in the way. In full consciousness, Beck believed that individuals are able to identify their limiting beliefs and challenge their typical habits, ways of thinking, and patterns of behavioral responses (Schenck, 2011), thus becoming a lifelong learner who continually expands their intellectual horizons.

Core beliefs, or schemas, are deeply held beliefs that people have about themselves, others, and the world. In other words, core beliefs act like a lens through which every situation and life experience is seen. Core beliefs are learned and influenced by early childhood experiences as people try to make sense of what is happening around them or to them. As they try to find meaning in these experiences, they develop a set of beliefs that are so fundamental and deep that they carry them into adulthood. These can include beliefs such as:

- "I'm worthless and hate myself." (self)
- "No one cares about me." (others)
- "It's pointless. Things will never get better." (world)

Core beliefs are not necessarily based on truths but on messages that individuals have received from others around them—such as family, friends, teachers, peers, and the media—which can influence their worldview in positive or negative ways.

Finally, *automatic thoughts* are thoughts that arise involuntarily in response to a person's day-to-day circumstances. When clients become overly focused on the negative, these thoughts can become negatively skewed, leading them to develop distorted perceptions of reality. It can be hard to recognize negative automatic thoughts because they pop up without effort and are fleeting, but you can help clients recognize these thoughts—instruct them to notice when they are experiencing distressing emotions, paying close attention to their internal dialogue in these moments.

Importantly, negative automatic thoughts fall into different categories of thinking errors, which are known as cognitive distortions. Helping clients understand the cognitive distortions they hold will help them become aware of negative thinking traps that contribute to higher levels of anxiety. They can then practice challenging and reframing these intrusive thoughts.

THERAPEUTIC WORKSHEET

My Core Beliefs

Core beliefs are deeply rooted ideas you have about yourself, others, and the world. When you have *rational* core beliefs, it leads to balanced reactions, but when you have *harmful* core beliefs, it leads to negative thoughts, feelings, and behaviors. Because everyone has different life experiences, two people in the same situation may think, feel, and behave very differently and have two completely different core beliefs. Here is an example:

Situation: Person A and Person B each go on a date. Neither of them is invited out for a second date.

Person A

Core belief: "I'm not good enough. I am unlovable."

Thought: "Of course he doesn't want to see me again. I'm not skinny enough and I probably said something stupid."

Feeling: Anxious, depressed

Behavior: Person A becomes obsessed with working out and going to the gym. They decide they will not start dating again until they look "perfect."

Person B

Core belief: "I am good enough to be loved by others. I am enough just as I am."

Thought: "I had a fun night, and I am glad I was able to have a nice conversation. I am going to keep putting myself out there because I know I will find the right person for me one day."

Feeling: Disappointed but hopeful

Behavior: Person B plans a night out with their friends, focuses on their own self-care, and continues to go on dates with other people.

To begin challenging your negative core beliefs, you must first identify what they are. Place a check mark next to any core beliefs that you identify with.

- ☐ I'm unlovable.
- ☐ I'm not good enough.
- ☐ I'm ugly.
- ☐ I'm worthless.
- ☐ I'm undeserving.
- ☐ I'm stupid.
- ☐ The world is dangerous.
- ☐ People can't be trusted.
- ☐ Everyone is out to get me.
- ☐ No one understands me.
- ☐ Nothing ever goes right.
- ☐ Life is unfair.
- ☐ I'm a failure.
- ☐ I don't have any purpose.
- ☐ I will never be successful.
- ☐ I'm never going to be happy.
- ☐ I'm too busy.
- ☐ I'm too young.
- ☐ I'm too old.
- ☐ I'm not smart enough.
- ☐ I will never find anyone to share my life with.
- ☐ Everything is my fault.
- ☐ Other: _____

PSYCHOEDUCATIONAL HANDOUT

Cognitive Distortions

Your thoughts can dictate your mood. It is important to remember this because managing your thoughts in a more rational way will help you feel better and ultimately improve your well-being. Once you understand your thought patterns more clearly, you can work on changing your mood by first changing your thoughts. There are a number of irrational thought patterns called *cognitive distortions*. These thought patterns can present in any anxiety-provoking situation.

Catastrophizing: You believe that the worst possible outcome will happen.

- "Another breakup! No one is ever going to love me, and I'm going to die alone."
- "I did horrible in that interview and will never get another job."

Overgeneralizing: You draw a conclusion based on a couple of experiences.

- "I failed this test, so I will never get into college."
- "No one has asked me on a date, so I will be alone forever."

Labeling: Labeling is an extreme form of overgeneralization in which you assign judgment to yourself or others based on one negative occurrence. Instead of recognizing that you or others made a mistake, you attach a label to it. This mislabeling of the situation is generally exaggerated and is solely based on that single incident.

- Your coworker is swamped with work, which causes him to be irritable and dismissive. You think, "I asked my coworker for help with something, and he totally dismissed me. What a jerk."

Personalization: You attribute a disproportionate amount of the blame for negative feelings or events to yourself. You fail to see that certain events are also caused by others.

- "My mom is always upset. She would be fine if I did more to help her."

Blaming: You wrongly blame others for your problems and play the victim role by holding others responsible for your pain.

- "My teacher refused to accept my homework that was late by only one day! He is *way* too strict."
- "My mom is the reason I am an addict."

Jumping to conclusions: You make unhelpful assumptions about a situation with little or no evidence.

- "That group of people are laughing at something—it must be me."
- "My dad is late to pick me up—something terrible must have happened."

Mind reading: You believe that you know what other people are thinking with no sufficient evidence.

- "She said no when I asked her out. She probably thinks I'm ugly."
- "Even though my friend told me they weren't upset, I know they are still angry with me."

Fortune telling: You expect that a situation will turn out badly without adequate evidence.

- "I will be alone forever."
- "Everyone at the party will think I'm a loser."

Emotional reasoning: You treat your emotions as facts and let your feelings guide your interpretation of reality.

- "I feel scared, therefore I am in danger."
- "I feel lonely, therefore no one loves me."

Disqualifying the positive: You recognize only the negative aspects of a situation while ignoring the positive.

- "Anyone could have written that book. I still haven't really achieved much."
- "I only went to the gym twice this week."

Should statements: You believe that things must be a certain way and focus on the way things "should" be instead of focusing on what is.

- "I should invite my sister to dinner or she will be mad at me."
- "I should be married by now. I am unlovable."

All-or-nothing thinking: You think in absolutes such as *always, never,* or *every*.

- "I'll never be good enough."
- "I always say the wrong thing."

What-ifs: You ask a series of questions about "what if" something happens and are never satisfied with any of the answers.

- "What if I get anxious?"
- "What if I can't catch my breath?"

THERAPEUTIC WORKSHEET

Cognitive Distortions Checklist

Review this list of cognitive distortions and place a check mark by the ones that you identify with. It can be one, a few, or all of them.

- ☐ **Catastrophizing:** Imagining and believing that the worst possible outcome will happen
- ☐ **Overgeneralizing:** Drawing a conclusion based on a couple of experiences
- ☐ **Labeling:** Assigning judgment to yourself or others based on one negative occurrence
- ☐ **Personalization:** Attributing a disproportionate amount of the blame for negative feelings or events to yourself
- ☐ **Blaming:** Wrongly blaming others for your problems and playing the victim role by holding others responsible for your pain
- ☐ **Jumping to conclusions:** Making unhelpful assumptions about a situation with little or no evidence
- ☐ **Mind reading:** Believing that you know what other people are thinking with no sufficient evidence
- ☐ **Fortune telling:** Expecting a situation will turn out badly without adequate evidence
- ☐ **Emotional reasoning:** Treating your emotions as facts and letting your feelings guide your interpretation of reality
- ☐ **Disqualifying the positive:** Recognizing only the negative aspects of a situation while ignoring the positive
- ☐ **Should statements:** Believing that things must be a certain way and focusing on the way things "should" be instead of focusing on what is
- ☐ **All-or-nothing thinking:** Thinking in absolutes such as *always*, *never*, or *every*
- ☐ **What-ifs:** Asking a series of questions about "what if" something happens and never being satisfied with any of the answers

Of the cognitive distortions you checked, identify the three you struggle with the most and give an example of when you had this type of thinking error.

1. _____

   Example: _____
   _____
   _____
   _____
   _____
   _____

2. _____

   Example: _____
   _____
   _____
   _____
   _____
   _____

3. _____

   Example: _____
   _____
   _____
   _____
   _____
   _____

Facts versus Opinions

In order to challenge a client's negative automatic thoughts, it's important for them to recognize the difference between facts and opinions. Although anxiety leads many people to believe that thoughts represent the absolute truth, the reality is that thoughts are not facts. They are simply a representation of our perspective at that particular moment. This can be a difficult idea to accept at first, especially when deep-rooted emotions arise, as our thoughts can have very powerful effects on how we feel and what we do.

We have over 50,000 thoughts a day, which can feel like an overwhelming number, especially for clients who feel as though their brains just won't shut off. If your clients are able to view each thought as a passing mental event rather than a reflection of fact, they may find they are able to deal with them in a more skillful way. This can then stop the vicious cycle of rumination. The following worksheet will help your clients see that while we all have a lot of emotionally charged automatic thoughts, they are not all absolute truths.

THERAPEUTIC WORKSHEET

Facts versus Opinions

Although we tend to believe that every single thought we have is true, *thoughts are not facts*. While some thoughts may be factual (e.g., "I failed the test"), others are not (e.g., "I am dumb"). These non-factual thoughts are opinions. This worksheet is designed to help you practice differentiating between factual thoughts and opinions. Read through each statement and place a check mark in the appropriate column to indicate whether it is a fact or opinion.

	Fact	Opinion
1. I'm dumb.		
2. I am single.		
3. I wear glasses.		
4. I will be single forever.		
5. I'm unattractive.		
6. There's something wrong with me.		
7. I am alone.		
8. Nobody likes me.		
9. My hair is curly.		
10. I'm a bad person.		
11. I failed the exam.		
12. Nothing ever goes right.		
13. I will never be loved.		
14. He shouted at me.		
15. I'm not good enough.		

Answer key: 1. opinion, 2. fact, 3. fact, 4. opinion, 5. opinion, 6. opinion, 7. fact, 8. opinion, 9. fact, 10. opinion, 11. fact, 12. opinion, 13. opinion, 14. fact, 15. opinion.

Breaking the Anxiety Cycle Using CBT

Now that your clients have a clearer picture of the CBT model, they can start creating treatment goals and think about the steps they will take to reach these goals, ultimately creating their anxiety healing toolkit! Creating a healing toolkit is essential in helping your client drop their avoidance behaviors. These behaviors often developed as coping mechanisms because the client felt as if they had no other options for self-regulation. The toolkit you will find in the next section is made up of holistic strategies that will help your client manage their anxiety in healthier and more effective ways.

In chapter 7, you will guide clients in rewiring their "fear-based brain" by helping them build greater emotional awareness, identify anxious triggers, rate their anxiety levels, and restructure irrational thought patterns. In chapters 8 through 11, you'll find a collection of holistic tools that clients will learn to practice in and outside of session. Since CBT is considered the most successful treatment method for anxiety disorders, these holistic strategies are centered around a mindfulness-based CBT approach to treating anxious clients. By practicing these tools regularly, clients can develop the ability to self-soothe and reduce the intensity of anxious feelings. Additionally, having these techniques readily available in their toolkit gives them a sense of control and empowerment over their anxiety.

PART IV
The Anxiety Healing Toolkit

CHAPTER 7

Rewire the Anxious Brain

In this chapter, you'll find cognitive exercises that allow clients to rewire their anxious brain. This step-by-step process involves teaching clients to (1) experience and express their emotions, (2) identify their anxious triggers, (3) rate their anxiety in response to these triggers, and (4) identify and challenge their negative automatic thoughts. As you work through this chapter, you may also decide to have clients complete the mindfulness tools in chapter 8—the choice is yours. Some clients may prefer to stay within the chronological structure of this guide, while others will benefit from practicing the techniques provided in chapter 8 right away. Use your clinical judgment, as client needs will vary.

Step 1: Experiencing and Expressing Emotions

When it comes to anxiety, many clients tend to internalize their feelings, keeping them bottled up inside. However, this only serves to intensify the anxiety and can lead to a vicious cycle of worry and stress. By enhancing your client's emotional vocabulary and teaching the importance of open feeling expression, they can break free from this cycle and begin to heal. This emotional awareness will help your client relate to other people, regulate their response to adversity, and make healthy choices. Note that emotional awareness may come more easily to some clients than others for a number of reasons, including (but not limited to) history of trauma, emotional immaturity of their caregivers, or lack of education around emotion regulation. The following exercises will help improve their emotional awareness.

IN-SESSION PRACTICE

Brain Dump

A brain dump is the first step in helping clients organize their anxious thoughts and understand their irrational thought patterns more clearly, including which ones may be contributing to their anxiety. Use the following script to guide the client in completing a brain dump in session.

Clinician Script

Today you're going to practice doing a brain dump for five minutes. A brain dump is the act of writing down everything that is in your brain at that moment. You basically write down everything you are thinking about or can hear yourself saying.

This exercise does not require proper punctuation or grammar, just the free flow of your thoughts on paper. It can act as a way to get all of your thoughts down on paper and free up space in your brain. This can help declutter your mind and increase your self-awareness, which can decrease stress.

The brain dump will also help you organize your anxious thoughts and understand your irrational thought patterns more clearly. Once you're able to identify which thoughts are taking up the most mental space in your mind, you will be able to focus on how to challenge and reframe them.

Let's practice. I'll set a timer for five minutes, and you will write down everything that's in your mind right now.

BETWEEN-SESSION WORKSHEET

Brain Dump

Take a few minutes to complete a brain dump. Set a timer for 5–10 minutes and write down anything that is in your brain at the moment. There are no rules. Don't worry about punctuation or grammar—just write! For instance, you might write about:

- Tasks that you need to accomplish
- Thoughts about the brain dump exercise
- Ideas that you're excited about
- Worries that are lingering in your mind

Your brain dump may even look like this: "This exercise is so dumb! I don't even know what to write, so I guess I'll just continue to write about how I have nothing to write about . . ."

You can use the following space to start your brain dump, but if you want to complete more than one brain dump before your next session, feel free to write it down anywhere you can!

THERAPEUTIC WORKSHEET

Feelings List

When you struggle with anxiety, it is common to report feeling "scared" or "worried," but what you may not realize is that there could be a number of other emotions present. This feelings list will help you become more aware of your emotions throughout the day, especially during times of high anxiety. By having a large feeling vocabulary to express yourself, you can better communicate with others, regulate yourself in times of high anxiety, and make healthier reactions to adversity. Read the following list of feeling words and circle any that you've had in the last 24 hours.

Afraid	Determined	Homesick	Moody	Self-conscious
Agitated	Disappointed	Hopeful	Mortified	Shocked
Alarmed	Disbelieving	Hopeless	Nauseated	Smug
Amazed	Discomforted	Horrified	Nervous	Sorrowful
Angry	Discontent	Humiliated	Nostalgic	Stressed
Anguished	Disgusted	Hurt	Numb	Stubborn
Annoyed	Disheartened	Impatient	Optimistic	Stuck
Anxious	Dismayed	Indifferent	Outraged	Submissive
Apprehensive	Disoriented	Infuriated	Overwhelmed	Suffering
Ashamed	Distracted	Insecure	Panicked	Surprised
Assertive	Distressed	Insightful	Paranoid	Suspenseful
Astonished	Disturbed	Insulted	Positive	Suspicious
Baffled	Elated	Interested	Proud	Terrified
Bewildered	Embarrassed	Intrigued	Puzzled	Thankful
Bitter	Enlightened	Irritated	Rageful	Thrilled
Blissful	Exasperated	Isolated	Regretful	Tired
Bored	Excited	Jealous	Rejected	Uncertain
Calm	Focused	Joyful	Relaxed	Uneasy
Carefree	Frustrated	Kind	Relieved	Unhappy
Careless	Gloomy	Lazy	Reluctant	Unsettled
Confident	Grateful	Lonely	Remorseful	Unsure
Confused	Grieving	Longing	Resentful	Upset
Courageous	Grumpy	Loopy	Restless	Vulnerable
Cowardly	Guilty	Loving	Sad	Weak
Curious	Happy	Mad	Satisfied	Worried
Depressed	Hateful	Melancholic	Scared	Other:
Despairing	Helpless	Miserable	Self-confident	_____

BETWEEN-SESSION WORKSHEET

Daily Check-In: Emotional Awareness

Take a few minutes each day to write about how you feel and why. This daily check-in will help you build greater emotional awareness. If needed, you can use the *Feelings List* worksheet for a detailed list of possible emotions. Under the reflection section for each day, describe your day, including the things you did, the places you went, or the people you interacted with.

Day/Time	Emotions	Intensity (1–10)
Monday		
Reflection:		
Tuesday		
Reflection:		
Wednesday		
Reflection:		

Day/Time	Emotions	Intensity (1–10)
Thursday		

Reflection:

Friday		

Reflection:

Saturday		

Reflection:

Sunday		

Reflection:

IN-SESSION PRACTICE

I-Statements

One way your clients can practice becoming more attuned to their inner emotions is by using I-statements. This allows them to practice naming and using feeling words more often.

Clinician Script

When people feel they are being blamed (whether rightly or wrongly), it is common for them to respond in a defensive manner. I-statements are an effective and simple way to communicate your feelings and emotions in these situations. An I-statement takes responsibility for your own feelings, while calmly describing the problem. An I-statement follows this format: "I feel [*feeling word*] when [*situation*] because [*explanation*]." When expressing yourself with I-statements, it is important to pay attention to the tone of your voice—using a soft, gentle tone while describing how the other person's actions affect you. Here are some examples.

1. Instead of: "You are always coming home so late! It's so rude!"

    I-Statement: "I feel worried when you come home late because I can't fall asleep until I know you're home safe."

2. Instead of: "You never clean up after yourself and the place is always a mess."

    I-Statement: "I feel frustrated when I come home and the house is messy because physical clutter affects my mental health."

THERAPEUTIC WORKSHEET

I-Statements

To get into the habit of using I-statements, read each of the following scenarios and fill in the blanks with a corresponding I-statement response.

Scenario 1

You're at a restaurant waiting for a friend who is 20 minutes late. This has become a pattern for your friend, and at times, this friend has even canceled plans at the last minute.

I feel _____ when _____

_____

because _____

_____.

Scenario 2

You are telling your partner about your day but notice that they are not listening (e.g., looking around the room and texting while you're talking).

I feel _____ when _____

_____

because _____

_____.

Scenario 3

Your dad recently passed away, and when you try to ask your mom about her own health, she tells you to mind your own business.

I feel _____ when _____

_____

because _____

_____.

BETWEEN-SESSION WORKSHEET

I-Statements

Over the next week, write about two triggering situations that caused you to experience a big emotional reaction. Then write down how you could have used an I-statement to handle the situation. This will help build your emotional vocabulary and allow you to become more aware of the emotions that are connected to your triggers.

Situation 1

Describe what happened: _____
_____
_____

How could you have used an I-statement to respond?

I feel _____ when _____
_____

because _____
_____.

Situation 2

Describe what happened: _____
_____
_____

How could you have used an I-statement to respond?

I feel _____ when _____
_____

because _____
_____.

Step 2: Identifying Triggers

Triggers are events or situations, whether real or imagined, that make clients feel anxious, upset, or out of control. For clients with anxiety, common triggers include being in social situations, not getting enough sleep, being the center of attention, receiving criticism from others, or facing confrontation of any kind. While some of these triggers are preventable, many are unavoidable because they are based on external factors (e.g., receiving criticism from others). By helping clients gain a deeper understanding of the situations that activate their anxiety, they can better identify how their thoughts, behaviors, and emotions are connected, and importantly, learn how to manage and cope with these triggers as they arise. The following worksheets will help your clients notice the patterns that underlie their unique anxiety triggers.

THERAPEUTIC WORKSHEET

Identifying Triggers

Triggers are events or situations, real or imagined, that induce anxiety. Triggers can vary from person to person, meaning that what causes anxiety in one person may not have the same effect on another. Triggers can be anything from people, places, and situations to events or even specific memories or thoughts. When you know your unique triggers, you can challenge them and eventually learn how to manage them. This worksheet will help you do just that. Place a check mark next to all of the triggers that apply to you. You may also add your own to the list.

- ☐ Large crowds
- ☐ Physical, sexual, or emotional abuse or neglect
- ☐ Work
- ☐ Fear of failure
- ☐ Poor performance
- ☐ Mistakes
- ☐ Home environment
- ☐ Thoughts of the future
- ☐ Small spaces
- ☐ Fear of dying
- ☐ Animals
- ☐ Accidents
- ☐ Trauma
- ☐ Lack of sleep
- ☐ Trying new things
- ☐ Meeting new people
- ☐ Confrontation
- ☐ Finances
- ☐ Family issues
- ☐ Forgetting something
- ☐ Fear of being alone
- ☐ Fear of not being accepted
- ☐ Roller coasters
- ☐ Heights
- ☐ Maintaining conversations
- ☐ Illness
- ☐ Other: _____

Describe your three biggest triggers in detail.

1. _____
   _____

2. _____
   _____

3. _____
   _____

When was the last time each trigger affected you?

1. _____
   _____

2. _____
   _____

3. _____
   _____

How do you typically respond when you're exposed to these triggers? Do you avoid? Become angry, anxious, or depressed?

_____
_____
_____
_____
_____

What thoughts do you have when you think of confronting these triggers?

_____
_____
_____
_____
_____

BETWEEN-SESSION WORKSHEET

Trigger Deep Dive

Just about anything can be a trigger. When you can recognize all the different factors that cause you to feel emotionally overwhelmed, you can start taking proactive steps to manage how you respond to them in the future. To begin, take your time to think about what may be a trigger for you in the following categories. For example, is there a specific emotion that triggers anxiety? How about a person or place? Or a certain activity?

Emotions (e.g., fear, hurt, sadness, anger, worry)

_____
_____
_____

Places (e.g., crowded areas, small spaces)

_____
_____
_____

Habits (e.g., drinking coffee, taking certain medications, not getting enough sleep)

_____
_____
_____

People (e.g., parents, people who don't like you)

_____
_____
_____

Thoughts (e.g., "I am too much," "I am unlovable")

_____

_____

_____

Activities (e.g., school, sporting events, parties, meetings)

_____

_____

_____

Times of year (e.g., holidays, birthdays)

_____

_____

_____

Other

_____

_____

_____

BETWEEN-SESSION WORKSHEET

Daily Check-In: Triggers

This worksheet will help you become more aware of what is triggering you when you are having an anxiety episode. If needed, you can use the checklist from the *Identifying Triggers* worksheet for some examples of common triggers. If you're able to remember the events leading up to your trigger, write them in the appropriate box.

Day/Time	Trigger	Events Before Trigger
Monday		
Tuesday		
Wednesday		
Thursday		
Friday		
Saturday		
Sunday		

Step 3: The Anxiety Rating Scale

Another key aspect of helping clients work through their anxiety involves teaching them to become more familiar with the intensity of their symptoms when confronted with anxiety-provoking stimuli. You can accomplish this by using the following anxiety rating scale. By instructing clients to rate their anxiety levels on a daily basis, they can better understand what events or situations create anxiety, giving them a reference point on when exactly to begin using their healing tools. Once clients complete their healing toolkit, the right-hand column of the scale gives them insight into the type of healing tools that will be most helpful in that moment, as well as the frequency at which they should practice these tools. The anxiety rating scale is also an important measurement tool that clients should use when practicing the healing techniques in chapters 8 through 11. The scale can assess the effectiveness of each specific coping tool as clients rate their anxiety before and after practicing each tool in session.

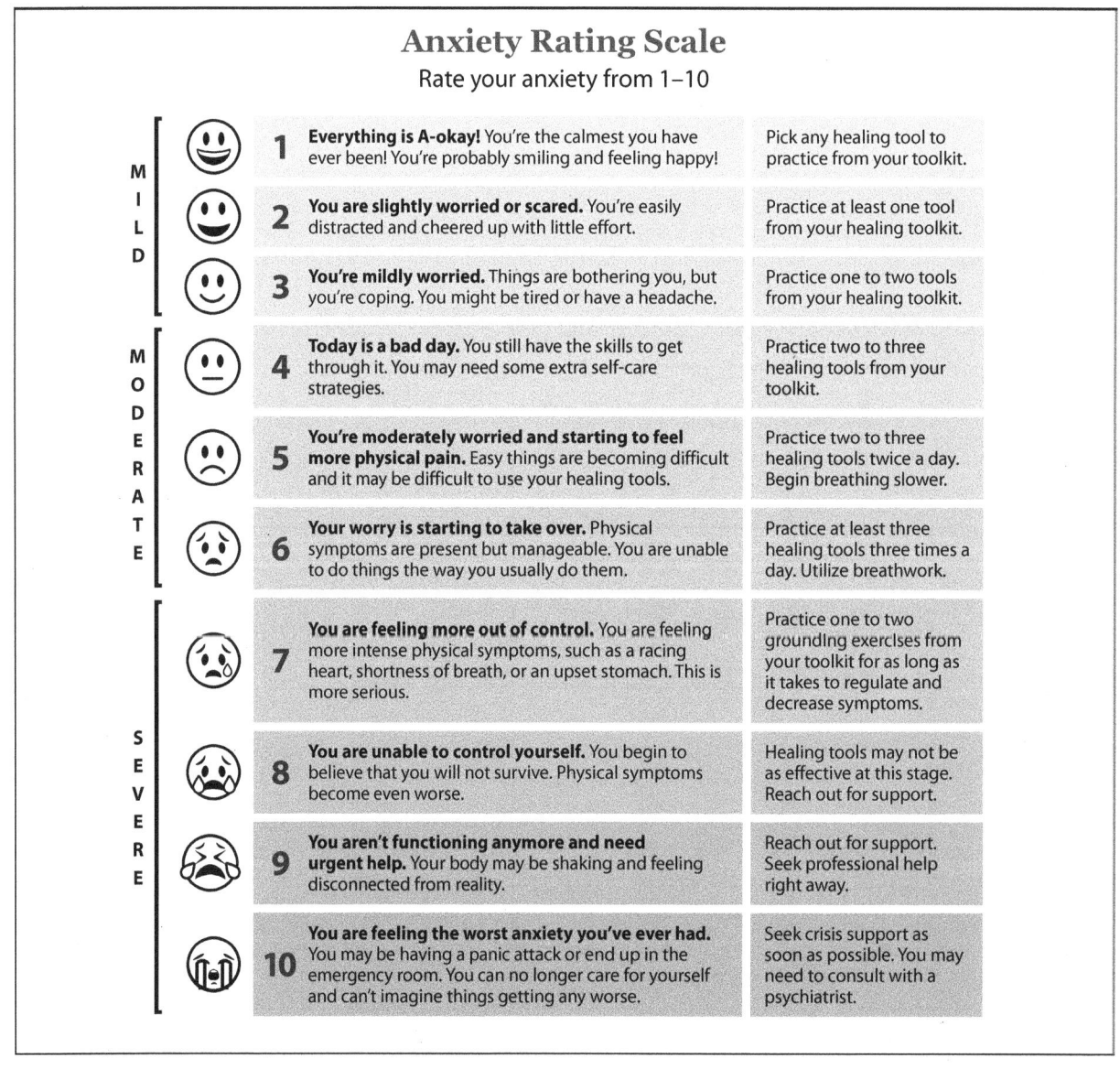

IN-SESSION PRACTICE

Rating Anxiety

The anxiety rating scale will help clients become more aware of when anxiety begins to escalate. This gives them an opportunity to step back and observe their thoughts and feelings as they unfold, which is the first step toward change and growth!

Clinician Script

To become more familiar with your triggers and patterns of anxiety, it is important to rate how intense your anxiety is each day. This will help you become more familiar with the events or situations that create anxiety and give you a reference point on what healing tools will be most effective in calming your nervous system. This rating scale will be an integral part of your healing toolkit. The following case study is one example of using the rating scale in everyday life.

- - - -

Sarah woke up at 3:30 a.m. and couldn't get back to sleep. She had to be up for work at 7:00 a.m. Sarah knew she always felt irritable and anxious when she didn't sleep well, and she also had a huge presentation at work that day. She got out of bed at 6:30 a.m. and immediately rated her anxiety at a 4. She knew that she would soon be a 5 or 6 once she got into her car and started driving to work. Sarah decided to dive into her healing toolkit and practice a couple strategies. Sarah completed a 10-minute guided meditation specifically tailored for work stress relief and reminded herself to practice square breathing. After this, her anxiety went down to a 3, and she felt ready to take on the workday. Sarah also continued to practice deep breathing and took several breaks throughout the day to get some fresh air. Before the presentation, Sarah practiced positive self-talk and said to herself in the bathroom mirror, "You got this, and if you make a mistake, don't worry. I still love you, and you are a rockstar." Sarah's anxiety never went above a 4 the rest of the day.

In the next chapter, you will learn specific tools to calm your anxiety, but for now you'll start getting into the habit of using this rating scale to recognize the intensity of your anxiety symptoms and notice when the rating begins to escalate.

THERAPEUTIC WORKSHEET

The Anxiety Rating Scale

This worksheet will help you become more aware of when anxiety begins to escalate. The anxiety rating scale will give you an opportunity to step back and observe your thoughts and feelings as they unfold, which is the first step toward change and growth. The more you become aware of what situations and environments trigger your anxiety, the more control you will have over how long it lasts!

Rate your anxiety from 1–10

MILD

1. **Everything is A-okay!** You're the calmest you have ever been! You're probably smiling and feeling happy! — Pick any healing tool to practice from your toolkit.

2. **You are slightly worried or scared.** You're easily distracted and cheered up with little effort. — Practice at least one tool from your healing toolkit.

3. **You're mildly worried.** Things are bothering you, but you're coping. You might be tired or have a headache. — Practice one to two tools from your healing toolkit.

MODERATE

4. **Today is a bad day.** You still have the skills to get through it. You may need some extra self-care strategies. — Practice two to three healing tools from your toolkit.

5. **You're moderately worried and starting to feel more physical pain.** Easy things are becoming difficult and it may be difficult to use your healing tools. — Practice two to three healing tools twice a day. Begin breathing slower.

6. **Your worry is starting to take over.** Physical symptoms are present but manageable. You are unable to do things the way you usually do them. — Practice at least three healing tools three times a day. Utilize breathwork.

SEVERE

7. **You are feeling more out of control.** You are feeling more intense physical symptoms, such as a racing heart, shortness of breath, or an upset stomach. This is more serious. — Practice one to two grounding exercises from your toolkit for as long as it takes to regulate and decrease symptoms.

8. **You are unable to control yourself.** You begin to believe that you will not survive. Physical symptoms become even worse. — Healing tools may not be as effective at this stage. Reach out for support.

9. **You aren't functioning anymore and need urgent help.** Your body may be shaking and feeling disconnected from reality. — Reach out for support. Seek professional help right away.

10. **You are feeling the worst anxiety you've ever had.** You may be having a panic attack or end up in the emergency room. You can no longer care for yourself and can't imagine things getting any worse. — Seek crisis support as soon as possible. You may need to consult with a psychiatrist.

Copyright © 2024 Alison Seponara, *The Anxiety Healer's Guide for Clinicians*. All rights reserved.

Explain what happened the last time you felt like your anxiety was at a 6 or above. How did you feel?

_____

_____

_____

_____

_____

What triggers did you notice at this time?

_____

_____

_____

_____

_____

How did you handle the situation?

_____

_____

_____

_____

_____

BETWEEN-SESSION WORKSHEET

Daily Check-In: Rating Your Anxiety Level

This worksheet will help you become more familiar with what is happening in your body throughout the day in relation to your anxiety level. Complete the log over the course of the next week and rate your anxiety level each day. Do you notice any specific triggers tied to your anxiety level?

Day	Anxiety Rating (1–10)			Triggers
	Morning	Afternoon	Evening	
Monday				
Tuesday				
Wednesday				
Thursday				
Friday				
Saturday				
Sunday				

Step 4: Cognitive Restructuring

Cognitive restructuring is a core technique used in CBT that involves helping clients identify unhelpful thought patterns and replace them with more realistic and positive thoughts, ultimately leading to improved emotional well-being and reduced anxiety. The first step in cognitive restructuring is to use Socratic questioning to challenge the validity and accuracy of a client's thoughts. Socrates was a Greek philosopher who emphasized the importance of questioning as a way to explore complex ideas and uncover assumptions. He believed that by questioning everything, we could gain a deeper understanding of the world around us and challenge our own beliefs and assumptions. You can use Socrates's philosophy to encourage clients to question the evidence and assumptions behind their anxious thoughts (Carey & Mullan, 2004). By taking a closer look at the evidence, clients can begin to see the flaws in their thinking.

This technique is simple. Once you have identified a client's cognitive distortion, encourage them to ask themselves the following series of questions:

- Is this thought realistic?
- Are you viewing this situation as black and white when it's really more complicated?
- Are you having this thought out of habit, or do the facts support it?
- Are you basing your thoughts on facts or on feelings?
- What evidence do you have to support this thought?
- What evidence do you have that contradicts this thought?
- Could you be misinterpreting the evidence?

Often, clients will find that there is very little evidence to support their worst-case scenario thought process. In fact, they might even find evidence to the contrary. Consider this example of challenging a client's automatic negative thought by looking at the factual evidence.

Examining the Evidence

Automatic Negative Thought: "I'm going to fail this presentation."

Therapist: What evidence do you have that supports this thought?"

Client: Well, I had a presentation three years ago and I felt like it went horribly. I stumbled on my words and didn't even get to the last slide because I ran out of time.

Therapist: What evidence do you have that you will 100 percent fail *this* presentation?

Client: I don't really know 100 percent for sure.

> **Therapist:** Why not?
>
> **Client:** Well, because I was given great feedback from my boss that I did a great job on the last one, and I do feel more prepared for this one.
>
> This series of questions highlights how the client's fear of failure is based on speculation rather than concrete facts.

For difficult cognitive distortions, you can also use the decatastrophizing technique, which involves asking a series of what-if questions to help clients consider how they would cope if their most feared outcome did, in fact, come true. By engaging in decatastrophizing, individuals can learn to approach their fears and worries from a more rational and balanced perspective. Let's take a look at an example to better understand how decatastrophizing works.

> ## Decatastrophizing
>
> ***Automatic Negative Thought: "This first date is going to end in disaster! I'm either going to embarrass myself or get rejected."***
>
> **Client:** I'm worried I will say something stupid and my date will think I'm boring. What if I make a fool of myself? What if they don't like me and never want to see me again?
>
> **Therapist:** So, what if those things come true? What if your date doesn't like you, or you make a fool of yourself? What's the worst that could happen?
>
> **Client:** Well, we probably won't have a second date.
>
> **Therapist:** What if you don't have a second date? What happens then?
>
> **Client:** I guess nothing. I just won't see them again.
>
> **Therapist:** What if you don't see them again?
>
> **Client:** I guess nothing. I will just go on other dates and find someone who likes me.
>
> This sequence of questioning highlights the fact that even the worst-case scenario is manageable. This typically brings a sense of relief and slows the panic that clients may experience by expecting the worst of the worst.

These techniques are most effective when used repeatedly with clients whenever cognitive distortions are identified. With enough practice, your client's cognitive distortions will lessen and be replaced with new, balanced thoughts.

PSYCHOEDUCATIONAL WORKSHEET

Socratic Questioning

Socratic questioning is a method you can use to question the evidence and assumptions behind your automatic negative thoughts. When you take a closer look at the facts, you will often find that there is very little evidence to support your automatic negative thoughts. In fact, you might even find evidence to the contrary. The next time you have a negative automatic thought, write it down in the space provided. Then go through each of the Socratic questions to see whether or not the thought holds up.

Negative automatic thought: _____

Socratic questions to ask yourself:

- Is this thought realistic?
  _____

- Am I viewing this situation as black and white when it's really more complicated?
  _____

- Am I having this thought out of habit, or do the facts support it?
  _____

- Am I basing my thoughts on facts or on feelings?
  _____

- What evidence do I have to support this thought?
  _____

- What evidence do I have that contradicts this thought?
  _____

- Could I be misinterpreting the evidence?
  _____

BETWEEN-SESSION WORKSHEET

Thought Log

A thought log is an effective tool to recognize how your thoughts, feelings, and behaviors affect one another. With enough practice, it will become natural to identify these connections, which will give you the power to begin challenging your negative automatic thoughts in real time. Complete the following thought log by describing an experience that caused anxiety, making sure to provide as much detail as possible in each category.

Date/Time	Situation (Trigger) What was going on before the anxiety escalated?	Automatic Negative Thought What were you thinking at the time?	Emotions What were you feeling at the time?	Body Sensations What physical symptoms did you experience?	Behaviors How did you respond? Did it make you feel better?	Examining the Evidence Does the evidence support the thought?
		How strongly do you believe this thought (1–10)?				
		What cognitive distortion(s) are present?	How intense were these feelings on the anxiety rating scale (1–10)?			What is a more realistic or helpful thought instead?

Copyright © 2024 Alison Seponara, *The Anxiety Healer's Guide for Clinicians*. All rights reserved.

CHAPTER 8

Breathwork and Grounding Techniques

Now that your client has created the first part of their healing toolkit and is more emotionally, physically, and cognitively aware of their anxiety, it's time to practice techniques that can help them regulate their nervous system when it's stuck in a state of fight or flight. These next chapters will help your client decide what tools will be most helpful to include in their anxiety healing toolkit. By practicing these tools in session together and at home, they will have a clearer understanding of what works for them in moments of fear, worry, and anxiety.

Breathwork

When a client has a panic attack or an anxiety attack, their sympathetic nervous system becomes activated, leading to a whole range of physical symptoms, including a rapid heart rate, heart palpitations, and shortness of breath. In order to return to a calm and balanced state, your client will need to practice strategies that activate the parasympathetic nervous system—especially deep, slow breathing. This produces a feeling of calmness and body connectedness that diverts attention away from stressful, anxious thoughts and quiets the mind. The breathwork exercises included in this chapter are:

- Square breathing
- Lion's breath
- Diaphragmatic (belly) breathing
- Zen word breathing
- Bellows breathing
- 4-7-8 breathing
- Alternate nostril breathing

Remember, not all breathing practices will be appropriate for every client. You may want to review the breathing techniques before practicing and assess which ones may be most helpful for your client. It may also be useful to rate your client's anxiety levels before and after each exercise using the anxiety rating scale from chapter 7.

IN-SESSION PRACTICE

Square Breathing

Introduce this breathing technique to your client in session. With practice, they will be able to add it to their healing toolkit to help reduce their anxiety symptoms. Encourage your client to repeat this exercise for three rounds, making sure to rate their anxiety before and after the exercise.

Clinician Script

Square breathing is a technique used to help slow down your heart rate. This method helps divert your attention from the irrational thought patterns that create anxiety. While this may not be a long-term solution for thinking anxious thoughts, this breathing technique can at least help with shortness of breath. Now, let's try it.

- - - -

Find a comfortable position and close your eyes.

Breathe in through your nose for a count of four.

(*Pause*)

Hold the breath for a count of four.

(*Pause*)

Breathe out through your mouth for a count of four.

(*Pause*)

Hold the breath for a count of four.

BETWEEN-SESSION WORKSHEET

Daily Practice: Square Breathing

Use this worksheet to record your daily practice of square breathing. Schedule specific times during each day to practice—the recommended practice is three rounds, three times per day. If needed, you can set a timer in your smartphone to help remind you. Rate your anxiety level before and after each practice, and then describe how you feel afterward.

Monday		
Square Breath Round 1	**Square Breath Round 2**	**Square Breath Round 3**
Time:	Time:	Time:
Anxiety Rating (1–10):	**Anxiety Rating (1–10):**	**Anxiety Rating (1–10):**
Before:	Before:	Before:
After:	After:	After:
I feel . . .	I feel . . .	I feel . . .
Tuesday		
Square Breath Round 1	**Square Breath Round 2**	**Square Breath Round 3**
Time:	Time:	Time:
Anxiety Rating (1–10):	**Anxiety Rating (1–10):**	**Anxiety Rating (1–10):**
Before:	Before:	Before:
After:	After:	After:
I feel . . .	I feel . . .	I feel . . .
Wednesday		
Square Breath Round 1	**Square Breath Round 2**	**Square Breath Round 3**
Time:	Time:	Time:
Anxiety Rating (1–10):	**Anxiety Rating (1–10):**	**Anxiety Rating (1–10):**
Before:	Before:	Before:
After:	After:	After:
I feel . . .	I feel . . .	I feel . . .

Thursday		
Square Breath Round 1	**Square Breath Round 2**	**Square Breath Round 3**
Time:	Time:	Time:
Anxiety Rating (1–10):	**Anxiety Rating (1–10):**	**Anxiety Rating (1–10):**
Before:	Before:	Before:
After:	After:	After:
I feel . . .	I feel . . .	I feel . . .
Friday		
Square Breath Round 1	**Square Breath Round 2**	**Square Breath Round 3**
Time:	Time:	Time:
Anxiety Rating (1–10):	**Anxiety Rating (1–10):**	**Anxiety Rating (1–10):**
Before:	Before:	Before:
After:	After:	After:
I feel . . .	I feel . . .	I feel . . .
Saturday		
Square Breath Round 1	**Square Breath Round 2**	**Square Breath Round 3**
Time:	Time:	Time:
Anxiety Rating (1–10):	**Anxiety Rating (1–10):**	**Anxiety Rating (1–10):**
Before:	Before:	Before:
After:	After:	After:
I feel . . .	I feel . . .	I feel . . .
Sunday		
Square Breath Round 1	**Square Breath Round 2**	**Square Breath Round 3**
Time:	Time:	Time:
Anxiety Rating (1–10):	**Anxiety Rating (1–10):**	**Anxiety Rating (1–10):**
Before:	Before:	Before:
After:	After:	After:
I feel . . .	I feel . . .	I feel . . .

IN-SESSION PRACTICE

Lion's Breath

Introduce this breathing technique with your client in session. With practice, they will be able to add it to their healing toolkit to help reduce their anxiety symptoms. Encourage your client to repeat this exercise for three rounds, making sure to rate their anxiety before and after the exercise.

Clinician Script

Lion's breath helps alleviate stress by stimulating the muscles in your throat and upper chest, which enervates the vagus nerve. This involves exhaling forcefully through your mouth, as if you were roaring like a lion. Now, let's practice.

Gently close your eyes and sit in a comfortable position with your hands on your knees.

Stretch out your arms into a goal-post position and stretch your fingers wide.

Take a deep breath in through your nose.

(*Pause*)

As you exhale, focus on the middle of your forehead or the end of your nose. Open your mouth as wide as you can and stick your tongue out, stretching it down toward your chin as far as it will go and release the breath while making a *HAAAAAA* sound.

BETWEEN-SESSION WORKSHEET

Daily Practice: Lion's Breath

Use this worksheet to record your daily practice of lion's breath. Schedule specific times during each day to practice—the recommended practice is two to three rounds, three times per day. If needed, you can set a timer in your smartphone to help remind you. Rate your anxiety level before and after each practice, and then describe how you feel afterward.

Monday		
Lion's Breath Round 1	**Lion's Breath Round 2**	**Lion's Breath Round 3**
Time:	Time:	Time:
Anxiety Rating (1–10):	**Anxiety Rating (1–10):**	**Anxiety Rating (1–10):**
Before:	Before:	Before:
After:	After:	After:
I feel . . .	I feel . . .	I feel . . .

Tuesday		
Lion's Breath Round 1	**Lion's Breath Round 2**	**Lion's Breath Round 3**
Time:	Time:	Time:
Anxiety Rating (1–10):	**Anxiety Rating (1–10):**	**Anxiety Rating (1–10):**
Before:	Before:	Before:
After:	After:	After:
I feel . . .	I feel . . .	I feel . . .

Wednesday		
Lion's Breath Round 1	**Lion's Breath Round 2**	**Lion's Breath Round 3**
Time:	Time:	Time:
Anxiety Rating (1–10):	**Anxiety Rating (1–10):**	**Anxiety Rating (1–10):**
Before:	Before:	Before:
After:	After:	After:
I feel . . .	I feel . . .	I feel . . .

Thursday		
Lion's Breath Round 1	**Lion's Breath Round 2**	**Lion's Breath Round 3**
Time:	Time:	Time:
Anxiety Rating (1–10):	**Anxiety Rating (1–10):**	**Anxiety Rating (1–10):**
Before:	Before:	Before:
After:	After:	After:
I feel . . .	I feel . . .	I feel . . .

Friday		
Lion's Breath Round 1	**Lion's Breath Round 2**	**Lion's Breath Round 3**
Time:	Time:	Time:
Anxiety Rating (1–10):	**Anxiety Rating (1–10):**	**Anxiety Rating (1–10):**
Before:	Before:	Before:
After:	After:	After:
I feel . . .	I feel . . .	I feel . . .

Saturday		
Lion's Breath Round 1	**Lion's Breath Round 2**	**Lion's Breath Round 3**
Time:	Time:	Time:
Anxiety Rating (1–10):	**Anxiety Rating (1–10):**	**Anxiety Rating (1–10):**
Before:	Before:	Before:
After:	After:	After:
I feel . . .	I feel . . .	I feel . . .

Sunday		
Lion's Breath Round 1	**Lion's Breath Round 2**	**Lion's Breath Round 3**
Time:	Time:	Time:
Anxiety Rating (1–10):	**Anxiety Rating (1–10):**	**Anxiety Rating (1–10):**
Before:	Before:	Before:
After:	After:	After:
I feel . . .	I feel . . .	I feel . . .

IN-SESSION PRACTICE

Diaphragmatic (Belly) Breathing

Introduce this breathing technique to your client in session. With practice, they will be able to add it to their healing toolkit to help reduce their anxiety symptoms. Encourage your client to repeat this exercise for three rounds, making sure to rate their anxiety before and after the exercise.

Clinician Script

Diaphragmatic or "belly" breathing is a specific type of breathwork that activates the relaxation response and allows the respiratory system to function correctly. Let's practice.

- - - - -

Get into a comfortable position, relax your shoulders, and close your eyes.

Place one hand on your chest and the other on your stomach. This will allow you to feel your belly move as you breathe.

(Pause)

Breathe in slowly through your nose for four seconds so that your stomach rises and moves out against your hand. The hand on your chest should remain as still as possible.

(Pause)

You should feel the air move through your nostrils and down into your stomach, making your stomach expand.

(Pause)

As you exhale, press gently on your stomach and feel the belly fall for 4 . . . 3 . . . 2 . . . 1.

BETWEEN-SESSION WORKSHEET

Daily Practice: Diaphragmatic (Belly) Breathing

Use this worksheet to record your daily practice of diaphragmatic breathing. Schedule specific times each day to practice—the recommended practice is 10 to 15 minutes per day. Set a timer in your smartphone to help remind you. Rate your anxiety level before and after each practice, and then describe how you feel afterward.

Day/Time	Anxiety Rating (1–10)		I feel . . .
	Before	After	
Monday			
Tuesday			
Wednesday			
Thursday			
Friday			
Saturday			
Sunday			

IN-SESSION PRACTICE

Zen Word Breathing

Introduce this breathing technique to your client in session. With practice, they will be able to add it to their healing toolkit to help reduce their anxiety symptoms. Encourage your client to repeat this exercise for three rounds, making sure to rate their anxiety before and after the exercise.

Clinician Script

A common stress response is shallow breathing, in which there is minimal breath moving through the lungs. Zen word breathing helps train your body to breathe deeply by using your diaphragm, helping more breath reach the lungs. Now, let's practice.

- - - -

Begin by closing your eyes and thinking of a word that brings you peace and joy.

(*Pause*)

Place one hand on your stomach and one hand over your heart, then take a deep breath in through your nose and out through your mouth.

(*Pause*)

Focus on the breath as the belly rises and falls.

(*Pause*)

As you take another breath in, repeat this Zen word in your mind.

(*Pause*)

As you exhale, repeat this Zen word again in your mind.

(*Pause*)

Take another breath in, and as you exhale, watch your belly fall and slowly spell this Zen word to yourself.

THERAPEUTIC WORKSHEET

My Zen Words

Use this worksheet to write down any words that bring you peace and joy. You will choose one of these words for a daily Zen breathing exercise that you will practice between sessions.

_____
_____
_____
_____
_____
_____
_____
_____
_____
_____
_____
_____
_____
_____
_____

BETWEEN-SESSION WORKSHEET

Daily Practice: Zen Word Breathing

Use this worksheet to record your daily practice of Zen word breathing. Schedule specific times each day to practice—the recommended practice is two to three rounds, three times per day. If needed, you can set a timer in your smartphone to help remind you. Rate your anxiety level before and after each practice, and then describe how you feel afterward.

My Zen word is: _____

Monday		
Zen Word Round 1	**Zen Word Round 2**	**Zen Word Round 3**
Time:	Time:	Time:
Anxiety Rating (1–10):	**Anxiety Rating (1–10):**	**Anxiety Rating (1–10):**
Before:	Before:	Before:
After:	After:	After:
I feel . . .	I feel . . .	I feel . . .

Tuesday		
Zen Word Round 1	**Zen Word Round 2**	**Zen Word Round 3**
Time:	Time:	Time:
Anxiety Rating (1–10):	**Anxiety Rating (1–10):**	**Anxiety Rating (1–10):**
Before:	Before:	Before:
After:	After:	After:
I feel . . .	I feel . . .	I feel . . .

Wednesday		
Zen Word Round 1	**Zen Word Round 2**	**Zen Word Round 3**
Time:	Time:	Time:
Anxiety Rating (1–10):	**Anxiety Rating (1–10):**	**Anxiety Rating (1–10):**
Before:	Before:	Before:
After:	After:	After:
I feel . . .	I feel . . .	I feel . . .

Thursday		
Zen Word Round 1	**Zen Word Round 2**	**Zen Word Round 3**
Time:	Time:	Time:
Anxiety Rating (1–10):	**Anxiety Rating (1–10):**	**Anxiety Rating (1–10):**
Before:	Before:	Before:
After:	After:	After:
I feel . . .	I feel . . .	I feel . . .

Friday		
Zen Word Round 1	**Zen Word Round 2**	**Zen Word Round 3**
Time:	Time:	Time:
Anxiety Rating (1–10):	**Anxiety Rating (1–10):**	**Anxiety Rating (1–10):**
Before:	Before:	Before:
After:	After:	After:
I feel . . .	I feel . . .	I feel . . .

Saturday		
Zen Word Round 1	**Zen Word Round 2**	**Zen Word Round 3**
Time:	Time:	Time:
Anxiety Rating (1–10):	**Anxiety Rating (1–10):**	**Anxiety Rating (1–10):**
Before:	Before:	Before:
After:	After:	After:
I feel . . .	I feel . . .	I feel . . .

Sunday		
Zen Word Round 1	**Zen Word Round 2**	**Zen Word Round 3**
Time:	Time:	Time:
Anxiety Rating (1–10):	**Anxiety Rating (1–10):**	**Anxiety Rating (1–10):**
Before:	Before:	Before:
After:	After:	After:
I feel . . .	I feel . . .	I feel . . .

IN-SESSION PRACTICE

Bellows Breathing

Introduce this breathing technique to your client in session. With practice, they will be able to add it to their healing toolkit to help reduce their anxiety symptoms. Encourage your client to repeat this exercise for three rounds, making sure to rate their anxiety before and after the exercise.

As a note, do not practice bellows breathing if your client is pregnant or has an ulcer, hiatal hernia, chronic constipation, heart disease, high blood pressure, uncontrolled hypertension, epilepsy, seizures, or panic disorder. You should also avoid practicing bellows breath if it has been less than two hours since the client has eaten.

Clinician Script

Bellows breathing, or *bhastrika*, is a yogic breathing technique that consists of a series of active inhalations and exhalations. When done properly, this practice increases energy and alertness—you may even feel as energized as you do after a good workout! You should feel the effort at the back of the neck, the diaphragm, the chest, and the abdomen. Now, let's practice.

- - - -

Sit up tall and relax your shoulders. Take a few deep breaths in and out from your nose and expand your belly fully as you breathe.

(*Pause*)

While keeping your mouth closed but relaxed, inhale and exhale rapidly and forcefully through your nose only at the rate of one second per cycle. This is a noisy breathing exercise, but that is what makes it so healing.

We will practice one round of 10 bellows breaths and then pause and breathe normally for 15 to 30 seconds.

Your breaths in and out should be equal in duration, but as short as possible. Make sure the breath is coming from your diaphragm, and keep your head, neck, shoulders, and chest as still as you can while your belly moves in and out.

Make sure to listen to your body during the practice. If you feel lightheaded in any way, let me know and take a pause for a few minutes while breathing naturally. When the discomfort passes, we can try another round slower and with less intensity.

BETWEEN-SESSION WORKSHEET

Daily Practice: Bellows Breathing

Use this worksheet to record your daily practice of bellows breathing. It is recommended that you practice in the morning, before a workout, before a yoga class, or anytime you're looking for energy. Try practicing three rounds each time by using the following guidance.

- **Round 1:** 10 bellows breaths. Pause and breathe normally for 15 to 30 seconds.
- **Round 2:** 20 bellows breaths. Pause and breathe normally for another 30 seconds.
- **Round 3:** 30 bellows breaths. Take a break and breathe naturally, observing the sensations in your mind and body.

Rate your anxiety level before and after each practice, and then describe how you feel afterward.

Day/Time	Anxiety Rating (1–10)		I feel...
	Before	**After**	

Day/Time	Anxiety Rating (1–10)		I feel . . .
	Before	After	

Disclaimer: Do not practice bellows breathing if you are pregnant or have an ulcer, hiatal hernia, chronic constipation, heart disease, high blood pressure, uncontrolled hypertension, epilepsy, seizures, or panic disorder. Always practice on an empty stomach, and do not perform while driving.

IN-SESSION PRACTICE

4-7-8 Breathing

Introduce this breathing technique to your client in session. With practice, they will be able to add it to their healing toolkit to help reduce their anxiety symptoms. Encourage your client to repeat this exercise for three rounds, making sure to rate their anxiety before and after the exercise.

As a note, some people may find it hard to breathe in for seven seconds and may become lightheaded. If your client has difficulty, you can practice this exercise with a smaller number of counts using the same ratio. For example, if you use a 3-5-6 breathing pattern, it will have the same effect.

Clinician Script

4-7-8 breathing is a very relaxing practice that may actually help ease you to fall asleep. It involves inhaling for a count of four, holding for a count of seven, and exhaling for a count of eight. This may seem like a subtle exercise when you first try it, but with repetition and practice, it can help you to gain better control over your breathing. The best thing about this exercise is that it doesn't require any equipment, takes little time to do, and can be done anywhere. Now, let's practice.

- - - - -

Begin by placing the tip of your tongue just behind your upper front teeth and keep it there through the entire exercise. Now close your eyes and breathe naturally for two rounds.

(*Pause*)

Make sure your mouth is closed and inhale quietly through your nose for a count of four.

(*Pause*)

Hold your breath for a count of seven.

(*Pause*)

Exhale completely through your mouth, making a whoosh sound, to a count of eight.

(*Pause*)

Now take one regular breath in through your nose and out through your mouth.

BETWEEN-SESSION WORKSHEET

Daily Practice: 4-7-8 Breathing

Use this worksheet to record your daily practice of 4-7-8 breathing. The recommended practice is two to three rounds, twice per day. It is especially beneficial to do this as you are trying to fall asleep. If needed, you can set a timer in your smartphone to help remind you. Rate your anxiety level before and after each practice, and then describe how you feel afterward.

Monday	
4-7-8 Breathing Round 1	**4-7-8 Breathing Round 2**
Time:	Time:
Anxiety Rating (1–10):	**Anxiety Rating (1–10):**
Before:	Before:
After:	After:
I feel . . .	I feel . . .
Tuesday	
4-7-8 Breathing Round 1	**4-7-8 Breathing Round 2**
Time:	Time:
Anxiety Rating (1–10):	**Anxiety Rating (1–10):**
Before:	Before:
After:	After:
I feel . . .	I feel . . .
Wednesday	
4-7-8 Breathing Round 1	**4-7-8 Breathing Round 2**
Time:	Time:
Anxiety Rating (1–10):	**Anxiety Rating (1–10):**
Before:	Before:
After:	After:
I feel . . .	I feel . . .

Thursday	
4-7-8 Breathing Round 1	**4-7-8 Breathing Round 2**
Time:	Time:
Anxiety Rating (1–10):	**Anxiety Rating (1–10):**
Before:	Before:
After:	After:
I feel . . .	I feel . . .

Friday	
4-7-8 Breathing Round 1	**4-7-8 Breathing Round 2**
Time:	Time:
Anxiety Rating (1–10):	**Anxiety Rating (1–10):**
Before:	Before:
After:	After:
I feel . . .	I feel . . .

Saturday	
4-7-8 Breathing Round 1	**4-7-8 Breathing Round 2**
Time:	Time:
Anxiety Rating (1–10):	**Anxiety Rating (1–10):**
Before:	Before:
After:	After:
I feel . . .	I feel . . .

Sunday	
4-7-8 Breathing Round 1	**4-7-8 Breathing Round 2**
Time:	Time:
Anxiety Rating (1–10):	**Anxiety Rating (1–10):**
Before:	Before:
After:	After:
I feel . . .	I feel . . .

Note: If you feel a little lightheaded when you first breathe this way, do not worry—it will pass! Do not do more than four breaths at one time for the first month of practice. Do not perform 4-7-8 breathing while driving. Although you can do the exercise in any position, as you are learning, it is best to sit with your back straight.

IN-SESSION PRACTICE

Alternate Nostril Breathing

Introduce this breathing technique to your client in session. With practice, they will be able to add it to their healing toolkit to help reduce their anxiety symptoms. Encourage your client to repeat this breathing technique for up to five minutes, making sure to rate their anxiety before and after the exercise.

As a note, alternate nostril breathing is best done on an empty stomach. Don't practice if your client is sick or congested.

Clinician Script

Alternate nostril breathing is another technique used to activate the parasympathetic nervous system and calm the mind and body. It is also a great way to enhance respiratory functioning and promote a feeling of balance. Now, let's practice.

- - - - -

Place your right thumb gently onto your right nostril, keeping the left nostril open.

Inhale slowly and deeply through your left nostril.

Then use your right ring finger to close your left nostril, while lifting up your thumb to open the right nostril, and exhale through your right nostril.

Inhale through your right nostril.

Then close your right nostril, while opening the left nostril, and exhale through your left nostril.

Now take two regular breaths.

BETWEEN-SESSION WORKSHEET

Daily Practice: Alternate Nostril Breathing

Use this worksheet to record your daily practice of alternate nostril breathing. Schedule specific times each day to practice—the recommended practice is three rounds, three times per day. It is especially beneficial to do this when you need to focus or relax. If needed, you can set a timer in your smartphone to help remind you. Rate your anxiety level before and after each practice, and then describe how you feel afterward.

Monday		
Alternate Nostril Round 1	**Alternate Nostril Round 2**	**Alternate Nostril Round 3**
Time:	Time:	Time:
Anxiety Rating (1–10):	**Anxiety Rating (1–10):**	**Anxiety Rating (1–10):**
Before:	Before:	Before:
After:	After:	After:
I feel . . .	I feel . . .	I feel . . .

Tuesday		
Alternate Nostril Round 1	**Alternate Nostril Round 2**	**Alternate Nostril Round 3**
Time:	Time:	Time:
Anxiety Rating (1–10):	**Anxiety Rating (1–10):**	**Anxiety Rating (1–10):**
Before:	Before:	Before:
After:	After:	After:
I feel . . .	I feel . . .	I feel . . .

Wednesday		
Alternate Nostril Round 1	**Alternate Nostril Round 2**	**Alternate Nostril Round 3**
Time:	Time:	Time:
Anxiety Rating (1–10):	**Anxiety Rating (1–10):**	**Anxiety Rating (1–10):**
Before:	Before:	Before:
After:	After:	After:
I feel . . .	I feel . . .	I feel . . .

Thursday		
Alternate Nostril Round 1	**Alternate Nostril Round 2**	**Alternate Nostril Round 3**
Time:	Time:	Time:
Anxiety Rating (1–10):	**Anxiety Rating (1–10):**	**Anxiety Rating (1–10):**
Before:	Before:	Before:
After:	After:	After:
I feel . . .	I feel . . .	I feel . . .

Friday		
Alternate Nostril Round 1	**Alternate Nostril Round 2**	**Alternate Nostril Round 3**
Time:	Time:	Time:
Anxiety Rating (1–10):	**Anxiety Rating (1–10):**	**Anxiety Rating (1–10):**
Before:	Before:	Before:
After:	After:	After:
I feel . . .	I feel . . .	I feel . . .

Saturday		
Alternate Nostril Round 1	**Alternate Nostril Round 2**	**Alternate Nostril Round 3**
Time:	Time:	Time:
Anxiety Rating (1–10):	**Anxiety Rating (1–10):**	**Anxiety Rating (1–10):**
Before:	Before:	Before:
After:	After:	After:
I feel . . .	I feel . . .	I feel . . .

Sunday		
Alternate Nostril Round 1	**Alternate Nostril Round 2**	**Alternate Nostril Round 3**
Time:	Time:	Time:
Anxiety Rating (1–10):	**Anxiety Rating (1–10):**	**Anxiety Rating (1–10):**
Before:	Before:	Before:
After:	After:	After:
I feel . . .	I feel . . .	I feel . . .

Disclaimer: Alternate nostril breathing is best done on an empty stomach. Don't practice if you're sick or congested.

Grounding Techniques

To find peace in the present moment, it is important that clients learn how to use grounding techniques that bring them back into the here and now in a safe way. This will allow them to remain present while struggling with intrusive and ruminating anxious thoughts. The more present they are in their body, the calmer and safer they will feel. Whether clients are on the go or feeling anxious at home, the grounding techniques provided here will help them remain more mindful when life feels overwhelming or too difficult to handle.

In addition, this section includes a series of guided meditations that will expand clients' ability to rest and relax. Guided meditations typically involve guiding a client to bring their awareness to any thoughts, feelings, or physical sensations they are experiencing—and observing these feelings and sensations without judgment. Instead of responding or reacting to those thoughts or feelings, have the client aim to simply note them and let them go. A consistent meditation practice can help train the brain to focus more on the present moment and less on the fears of the future. It is recommended that you or the client record the meditations in session so they can listen and practice between sessions.

The grounding and calming techniques in this section include:

- Progressive muscle relaxation
- Guided meditations:
    - You are safe
    - Body scan
    - The breath
    - The five senses
    - Happy place
- Grounding action steps

IN-SESSION PRACTICE

Progressive Muscle Relaxation

Introduce this technique to your client in session. With practice, they will be able to add it to their healing toolkit to help reduce their anxiety symptoms. Encourage your client to rate their anxiety before and after the exercise.

Clinician Script

Progressive muscle relaxation is a calming exercise where you systematically tense and relax various muscle groups throughout your body. Without straining, forcefully tense each muscle group for five seconds and then suddenly release the tension and feel the muscle relax. If any pain or discomfort arises, skip to the next muscle group. Remember to keep breathing throughout this exercise to keep your body in a balanced state. Now, let's practice.

- - - -

Begin by settling into a comfortable position. Take a deep breath as you inhale through your nose and slowly exhale through your mouth.

Take another breath in through your nose, and feel your abdomen rise as your lungs fill with air.

Hold the breath at the top for a few seconds.

As you exhale, imagine tension releasing from your body and slowly melting away.

Your body is feeling more and more relaxed with every breath.

As we move through each muscle, remember to keep breathing. You may notice your mind begin to wander during this practice. This is normal and expected. If this happens, just bring yourself back to the moment, without judgment, and focus on the muscle you are working on.

Now, let's begin.

Begin by making a fist and tightening the muscles in your *hands*. Clench your fists as tight as you can for 5 . . . 4 . . . 3 . . . 2 . . . 1 . . . and release.

As you feel the muscles in your hands relax, take a deep breath in through your nose and slowly exhale through your mouth.

(*Pause*)

Next, extend your *arms* in front of you and bend your *wrists* to point your fingers to the ceiling. Hold this tight for 5 . . . 4 . . . 3 . . . 2 . . . 1 . . . and release.

As you feel the muscles in your arms and wrists relax, take a deep breath and picture the stress leaving your body.

(*Pause*)

Now, let's move to the *biceps and upper arms*. Make a fist and bend your arms at the elbows while flexing your biceps. Hold the tension for 5 . . . 4 . . . 3 . . . 2 . . . 1 . . . and release.

As you feel the muscles in your biceps and upper arms relax, take a deep breath in through your nose and slowly exhale through your mouth. Your body is fully relaxed.

(*Pause*)

If your mind begins to wander, slowly bring your mind back to this moment. Right here. Right now. Your mind and body are letting go of all of your anxieties.

Next, shrug your *shoulders* up to your ears and tense the muscles tightly for 5 . . . 4 . . . 3 . . . 2 . . . 1 . . . and release.

As you feel the muscles in your shoulders relax, notice the difference between tension and relaxation. Take a deep breath in through your nose and notice your belly rise. Exhale slowly through your mouth and feel your belly fall.

(*Pause*)

Now moving to the face, tense the muscles in your *forehead* by raising your eyebrows as high as you can for 5 . . . 4 . . . 3 . . . 2 . . . 1 . . . and release.

As you feel the muscles in your forehead relax, take a deep breath in through your nose and slowly exhale through your mouth.

(*Pause*)

Now, close your *eyes* as tight as is comfortable and tense around the bridge of your nose for 5 . . . 4 . . . 3 . . . 2 . . . 1 . . . and release.

As you feel the muscles around your eyes relax, feel the tension float away as relaxation fills your body.

Take a deep breath in through your nose and slowly exhale through your mouth.

(*Pause*)

Next, smile as wide as you can, tensing the muscles around the *mouth and cheeks* for 5 . . . 4 . . . 3 . . . 2 . . . 1 . . . and release.

As you feel the muscles in your mouth and cheeks relax, take a deep breath in through your nose and slowly exhale through your mouth.

(*Pause*)

If your mind has wandered, you can always use your breath to bring yourself back to this moment.

Now slowly move your head back toward your back, as if you were looking up at the sky, and tense the *back of the neck*. Gently hold for 5 . . . 4 . . . 3 . . . 2 . . . 1 . . . and release.

Take a slow, deep breath in and slowly exhale all of your worries.

(*Pause*)

Now slowly touch your chin to your chest, tensing the *front of the neck* . . . hold for 5 . . . 4 . . . 3 . . . 2 . . . 1 . . . and release.

As you feel the muscles in your neck relax, take another breath in and visualize a white light full of positive energy. Slowly exhale and feel your body at ease.

(*Pause*)

Now moving down to your *chest*, take a deep breath in and hold for 5 . . . 4 . . . 3 . . . 2 . . . 1 . . . and release. Exhale, noticing any sensations that arise in your body, and allow your body to let go of any tension.

(*Pause*)

Next, arch your *back* away from the floor or chair and hold for 5 . . . 4 . . . 3 . . . 2 . . . 1 . . . and release. As you feel your body become limp, take a deep breath in through your nose and slowly exhale through your mouth.

(*Pause*)

Now tense the muscles in your *stomach* and suck in tight like a knot and hold for 5 . . . 4 . . . 3 . . . 2 . . . 1 . . . and release.

As you feel the muscles in your stomach relax, take a deep breath in through your nose and slowly exhale through your mouth.

(*Pause*)

Now tense your *buttocks* by pressing your buttocks together tightly and hold 5 . . . 4 . . . 3 . . . 2 . . . 1 . . . and release.

As you feel the muscles in your buttocks relax, picture stress leaving your body as you take a slow breath in and out.

(*Pause*)

Next, tense your *thigh* muscles together tightly and hold for 5 . . . 4 . . . 3 . . . 2 . . . 1 . . . and release.

As you feel the muscles in your thighs go limp, breathe in slowly through your nose and slowly exhale through your mouth, letting go of the tension. Feel the weight of your legs sink into the chair or ground.

(*Pause*)

Now point your toes toward your face, tense your *calves*, and hold for 5 . . . 4 . . . 3 . . . 2 . . . 1 . . . and release.

As you feel the muscles in your toes and calves relax, breathe in peace and exhale any stress or negativity. Feel the body completely relax and surrender.

(*Pause*)

Now moving down the body, curl your toes downward away from your face and tense your *feet*. Hold for 5 . . . 4 . . . 3 . . . 2 . . . 1 . . . and release.

As you feel the muscles in your toes relax, gently breathe in through your nose and slowly exhale through your mouth. Feel the body completely relax and surrender to the breath.

(*Pause*)

Your entire body is now completely calm and at ease. Your mind has released any negative thoughts, and your body is filled with the warmth of tranquility.

(*Pause*)

Take another deep breath in and slowly exhale.

Breathe in . . . Breathe out . . .

BETWEEN-SESSION WORKSHEET

Daily Practice: Progressive Muscle Relaxation

Use this worksheet to record your daily practice of progressive muscle relaxation (PMR). Schedule specific times each day to practice—the recommended practice is once a day, but you may want to use this practice more often if you experience intense physical symptoms when anxious. If needed, you can set a timer in your smartphone to help remind you. Rate your anxiety level before and after each practice, and then describe how you feel afterward.

Day/Time	Anxiety Rating (1–10)		I feel . . .
	Before	**After**	
Monday			
Tuesday			
Wednesday			
Thursday			
Friday			
Saturday			
Sunday			

IN-SESSION PRACTICE

You Are Safe Meditation

Introduce this meditation to your client in session. With practice, they will be able to add it to their healing toolkit to help reduce their anxiety symptoms. Encourage your client to rate their anxiety before and after the exercise.

Clinician Script

When you're anxious or panicked, your thoughts may wander to the worst-case scenario, and you may focus on the worry, which fuels your sense of fear. This guided meditation will help you achieve a more relaxed state of mind in these moments. It involves focusing your awareness on your breath and noticing any thoughts, feelings, or physical sensations you are experiencing in that moment. You then observe those thoughts, feelings, and sensations without judgment, making note of them and letting them go instead of reacting to them. Now, let's begin.

- - - -

Close your eyes and take a deep breath in and out.

Breathe in through your nose . . . 2 . . . 3 . . . 4.

Then breathe out through your mouth . . . 2 . . . 3 . . . 4 . . . 5.

(*Pause*)

Remember that you are safe in this moment.

(*Pause*)

You are safe in this moment.

(*Pause*)

You are safe in this moment.

(*Pause*)

No matter where your thoughts take you . . . no matter what scary thoughts may come to your mind . . . remember, you are safe.

(*Pause*)

Take a deep breath in for a count of four . . . hold it for two . . . and then breathe out for five.

(*Pause*)

When a scary thought pops into your head, remember that it is just a thought. The thought has no power. Simply watch the thought as if it were a cloud floating by.

(*Pause*)

Observe the thought . . . wave to the thought . . . and watch it float away.

(*Pause*)

And remember, you are safe.

(*Pause*)

Breathe in through your nose . . . 2 . . . 3 . . . 4.

Hold . . . 2 . . . 3.

Then breathe out through your mouth . . . 2 . . . 3 . . . 4 . . . 5.

Breathe in through your nose . . . 2 . . . 3 . . . 4.

Hold . . . 2 . . . 3.

Then breathe out through your mouth . . . 2 . . . 3 . . . 4 . . . 5.

Now as you relax, you can count your breaths as they continue to flow gently. Count ten breaths.

(*Pause*)

Notice the breath as it enters your nose, feeling it pass through your nasal passages and down your throat. And as you exhale through your mouth, notice as the air leaves your lungs.

(*Pause*)

Observe how the breath flows slowly.

(*Pause*)

You are safe. You are safe in this moment. Notice the calm state of your mind. Notice that your body is relaxed and at peace.

(*Pause*)

Now begin to bring yourself back to the present moment and into your usual level of alertness and awareness. Wiggle your toes and fingers and feel the surface beneath you. Listen for the sounds around you as you become fully alert.

When you are ready, open your eyes and take a moment to stretch your muscles and allow your body to reawaken. Move forward with your day as you feel refreshed, energized, calm, and relaxed.

IN-SESSION PRACTICE

Body Scan Meditation

Introduce this meditation to your client in session. With practice, they will be able to add it to their healing toolkit to help reduce their anxiety symptoms. Encourage your client to rate their anxiety before and after the exercise.

Clinician Script

A body scan meditation is an effective technique that can help reduce anxiety and promote relaxation. This practice involves bringing your attention to different parts of your body one by one while noticing any sensations or tension that may be present. By consciously focusing on each part of your body, you can become more aware of any areas of tension or discomfort and work toward releasing them. Now, let's get started.

- - - - -

Find a comfortable seated position and begin this meditation by taking a deep breath in through your nose for 2 . . . 3 . . . 4 . . . and out through your mouth for 2 . . . 3 . . . 4 . . . 5 . . . 6.

Take this time to close your eyes if you can and focus on your body sensations as you take another deep breath in for 2 . . . 3 . . . 4 . . . and out for 2 . . . 3 . . . 4 . . . 5 . . . 6.

(*Pause*)

As your body begins to relax and the stress starts to float away, roll your shoulders forward . . . and then back. Repeat this once more: Roll your shoulders forward—then back.

Focus on your body sensations and take another deep breath in through your nose for 2 . . . 3 . . . 4 . . . and out through your mouth for 2 . . . 3 . . . 4 . . . 5 . . . 6.

(*Pause*)

Next, stretch your arms out above your head, reaching your arms high, and spread your hands wide open above your head.

(*Pause*)

Now relax your hands and lower your arms.

Begin to focus on your body sensations and take another deep breath in through your nose for 2 . . . 3 . . . 4 . . . and out through your mouth for 2 . . . 3 . . . 4 . . . 5 . . . 6.

(*Pause*)

Now, let your shoulders relax and lower them away from your ears. Relax your jaw by dropping the lower jaw slightly and remove your tongue from the top of your mouth if it is there. Try to keep your top teeth from touching your bottom teeth.

Copyright © 2024 Alison Seponara, *The Anxiety Healer's Guide for Clinicians*. All rights reserved.

As you fall into a state of tranquility, feel your body become still, taking a deep breath in through your nose for 2 . . . 3 . . . 4 . . . and out through your mouth for 2 . . . 3 . . . 4 . . . 5 . . . 6.

(*Pause*)

Now repeat these words to yourself:

My body is safe right here . . . right now.
My body is safe right here . . . right now.
My body is safe right here . . . right now.

(*Pause*)

I am stronger than I think, and I will get through this.
I am stronger than I think, and I will get through this.
I am stronger than I think, and I will get through this.

(*Pause*)

Take a deep breath in for 2 . . . 3 . . . 4 . . . and out for 2 . . . 3 . . . 4 . . . 5 . . . 6.

(*Pause*)

I believe in myself, and I believe in my breath.
I believe in myself, and I believe in my breath.
I believe in myself, and I believe in my breath.

(*Pause*)

All will be well. Just breathe.
All will be well. Just breathe.
All will be well. You need this. Just breathe.

(*Pause*)

Feel the stillness and peace of this moment.

Right now . . . my body is love . . . my body is at peace . . . my body is at rest.

(*Pause*)

Take one more deep breath in through your nose for 2 . . . 3 . . . 4 . . . and out through your mouth for 2 . . . 3 . . . 4 . . . 5 . . . 6.

Now begin to bring yourself back to the present moment and into your usual level of alertness and awareness. Wiggle your toes and fingers and feel the surface beneath you. Listen for the sounds around you as you become fully alert.

When you are ready, open your eyes and take a moment to stretch your muscles and allow your body to reawaken. Move forward with your day as you feel refreshed, energized, calm, and relaxed.

IN-SESSION PRACTICE

The Breath Meditation

Introduce this meditation to your client in session. With practice, they will be able to add it to their healing toolkit to help reduce their anxiety symptoms. Encourage your client to rate their anxiety before and after the exercise.

Clinician Script

It's important to practice slowing your breath when you're anxious. Breathwork triggers the release of endorphins and other feel-good chemicals in your body, promoting a sense of calm while relaxing your muscles, relieving tension, and lowering your blood pressure. When you are anxious, your breath becomes shallow and rapid, which sends signals to your brain that you are in danger. By consciously slowing down your breath, you can send a message to your brain that everything is okay, which can help to regulate your nervous system. Now, let's get started.

– – – –

Take a moment and welcome yourself into this moment, allowing yourself to be in the here and now. Congratulate yourself for taking the time to be present and wander into a peaceful and safe space inside of your mind.

Begin by checking in with yourself and finding your breath. Take a deep breath in through your nose for a count of four . . . hold for two . . . then breathe out through your mouth for a count of seven.

(*Pause*)

Notice whether you feel any sensations or tightness in your body, as well as how your mood feels. Tune into your emotions and acknowledge whatever you are feeling—just letting it be. Whatever you are feeling at this very moment is valid. You are safe within these feelings. Find awareness and acknowledge whatever there is to be felt.

Now bring your attention to the breath. Be mindful of your breath as your abdomen expands with each inhale and falls with each exhale. You are safe in this moment. You are alive and breathing in this moment.

Take a slow breath in through your nose for four . . . and out through your mouth for seven.

(*Pause*)

As you breathe normally and naturally, feel the rise and fall of your stomach. You are safe in this moment. Your heart is slowing down and finding peace and comfort in this moment. If your mind has wandered off, compassionately and gently make a note of it "wandering" and then come back to the breath.

(*Pause*)

As you begin to get deeper into the stillness of the body and mind, you may experience some anxious thoughts, worries, or fears. If so, remember that they are just thoughts and they cannot hurt you. You are safe in this moment.

Choose to slow down and focus on your breath, taking one inhale and one exhale at a time. Now focus on your body and tap into any sensations, thoughts, or emotions you are experiencing. Whatever it is that you're feeling, acknowledge what is being felt and let it be.

(*Pause*)

Simply follow my voice, and if thoughts come up from time to time, do not worry. Simply return to the breath. Allow yourself to become steady and calm. Find relaxation in the whole body. Focus on the body and become aware of the importance of total stillness.

As you breathe in through your nose and out through your mouth, focus your awareness on the motion of your stomach. Concentrate on your stomach as it gently rises and falls. With every breath, feel your stomach expand. Breathing in . . . and breathing out.

(*Pause*)

Now bring the focus to your chest. With every breath you take, become aware of your chest gently rising and falling. Concentrate on the breath and continue to allow yourself to be at peace.

As you learn to be with things as they are, you will discover that your fears are truly just a figment of what is.

Right now, at this moment, nothing can hurt you. You are right where you are supposed to be.

(*Pause*)

Now begin to bring yourself back to the present moment and into your usual level of alertness and awareness. Wiggle your toes and fingers and feel the surface beneath you. Listen for the sounds around you as you become fully alert.

When you are ready, open your eyes and take a moment to stretch your muscles and allow your body to reawaken. Move forward with your day as you feel refreshed, energized, calm, and relaxed.

IN-SESSION PRACTICE

The Five Senses Meditation

Introduce this meditation to your client in session. With practice, they will be able to add it to their healing toolkit to help reduce their anxiety symptoms. Encourage your client to rate their anxiety before and after the exercise.

Clinician Script

Your senses—sight, hearing, touch, smell, taste—are your window to the world and how you experience life. This mindfulness exercise involves paying attention to one sense at a time in order to reconnect you with the present moment. When you're in panic mode, you lose the ability to think clearly, so bringing your mind to the things you can see, hear, feel, smell, and taste around you can bring you back to reality and create a calmer state, both mentally and physically. Plus, this technique can be used anywhere! Now, let's get started.

- - - - -

Begin this meditation by taking three deep breaths in through your nose . . . and out through your mouth.

Inhaling . . . 1 . . . 2 . . . 3 . . . 4 . . . and exhaling . . . 1 . . . 2 . . . 3 . . . 4 . . . 5 . . . 6.

Inhaling . . . 1 . . . 2 . . . 3 . . . 4 . . . and exhaling . . . 1 . . . 2 . . . 3 . . . 4 . . . 5 . . . 6.

Inhaling . . . 1 . . . 2 . . . 3 . . . 4 . . . and exhaling . . . 1 . . . 2 . . . 3 . . . 4 . . . 5 . . . 6.

(*Pause*)

Look around and begin using your sense of sight to focus on five things you can see. What are different shapes you notice? Look around you and silently describe the colors you notice.

(*Pause*)

Next, find four things around you that you can touch. What do you notice about the texture of what you can touch? Run your fingertips gently up the inside of your arm.

Feel the air across your skin. Bring awareness to your clothing and how it rests on your body. What is the texture of your pants and shirt? Hot, cold, rough, hard, soft—there are many sensations to experience.

(*Pause*)

Continue as you take a deep breath in . . . 2 . . . 3 . . . 4 . . . and out . . . 2 . . . 3 . . . 4 . . . 5 . . . 6.

(*Pause*)

Now close your eyes, if you can, and listen for three things you can hear. This can be any type of external sound. Maybe you hear cars driving by, your stomach growling, the wind blowing, or people talking. Whatever it is that you hear, focus on three things that you can hear outside of your body.

(*Pause*)

Take another deep breath in through your nose . . . and a long sigh out through your mouth.

(*Pause*)

Now focus on your sense of smell, and in your mind, describe two things you can smell. Maybe you can smell your perfume or cologne, a candle burning, your fresh clothes from the washer, or the outside air.

(*Pause*)

Now, let's turn to your sense of taste. What is one thing that you notice about the taste in your mouth? What does the inside of your mouth taste like? Gum, mint, coffee, or another flavor?

(*Pause*)

As you take a deep breath in . . . 2 . . . 3 . . . 4 . . . and out . . . 2 . . . 3 . . . 4 . . . 5 . . . 6. Remind yourself that you are safe right now . . . in this moment. You are alive, and you are breathing.

(*Pause*)

Now begin to bring yourself back to the present moment and into your usual level of alertness and awareness. Wiggle your toes and fingers and feel the surface beneath you. Listen for the sounds around you as you become fully alert.

When you are ready, open your eyes and take a moment to stretch your muscles and allow your body to reawaken. Move forward with your day as you feel refreshed, energized, calm, and relaxed.

BETWEEN-SESSION WORKSHEET

The Five Senses Meditation

Use this worksheet to complete the five senses meditation during highly anxious moments. Rate your anxiety before and after each practice, and then describe how you feel afterward.

Anxiety rating before (1–10): _____

Five things you can see:

1. _____
2. _____
3. _____
4. _____
5. _____

Four things you can touch:

1. _____
2. _____
3. _____
4. _____

Three things you can hear:

1. _____
2. _____
3. _____

Two things you can smell:

1. _____
2. _____

One thing you can taste:

1. _____

Anxiety rating after (1–10): _____

I feel . . .

_____

_____

_____

_____

_____

IN-SESSION PRACTICE

Happy Place Meditation

Introduce this meditation to your client in session. With practice, they will be able to add it to their healing toolkit to help reduce their anxiety symptoms. Encourage your client to rate their anxiety before and after the exercise.

Clinician Script

Visualization is a powerful tool that can help calm the mind and body. When you imagine yourself in a peaceful and happy environment, your brain responds by releasing serotonin, a neurotransmitter associated with positive emotions. Visualizing your happy place not only provides temporary relief from anxiety, but it can also help rewire your brain over time. By repeatedly envisioning yourself in a positive and calm environment, you can create new neural connections that strengthen the association between this mental image and relaxation. This may train your brain to respond more positively to anxiety-provoking situations, reducing the intensity and frequency of anxious episodes.

- - - - -

Begin this meditation by taking a deep breath in through your nose for a count of four and out through your mouth for a count of seven.

Become aware of your thoughts, watching them as they come in, then letting them go.

Your body is becoming relaxed, and your breath is becoming slower and calmer.

(*Pause*)

All you need to do right at this moment is breathe and listen . . . just breathe and listen. Focus on your breath as you breathe in and your stomach rises . . . and as you exhale, your stomach falls, deep and relaxed.

Feel your body relaxing deeper and deeper, and allow your thoughts to slow down, your limbs to become limp, and your mind to calm.

(*Pause*)

Now if you can, close your eyes. If you are unable to do so, that is totally okay. Notice your breath and take another deep breath in through your nose for a count of four . . . hold for three . . . and breathe out through your mouth for seven. Allow yourself to completely relax.

(*Pause*)

Now as you continue to breathe, use your imagination and picture a happy scene in your mind. Visualize yourself at this happy place. This could be somewhere you have been before or someplace that you create in your mind. This is a place full of inner peace and calm. It is completely anxiety-free.

(*Pause*)

As you visualize this happy place, breathe deep into your belly, letting your stomach rise up as you breathe in and fall as you release the breath.

Now picture yourself in this happy place, a warm and comforting setting that makes you smile. You are surrounded by all of the things that create so much joy and peace in your life. Take a few moments and create this scene in your mind. What do you see? What colors do you notice? Who is there with you in this place? Maybe it is just you, and that is wonderful.

(*Pause*)

Visualize that place and see it very clearly in your mind. It is a wonderful place, isn't it? You are happy. You are healthy. You are safe.

Take a deep breath in, and as you exhale, allow yourself to feel even more comfortable and relaxed. Take another slow breath in for four . . . and as you exhale, allow yourself to feel the joy and peace of your happy place. Allow yourself to feel completely relaxed. Completely at peace. Completely surrendering. Notice that you are in a safe place with no anxiety . . . and nothing can hurt you.

(*Pause*)

Now really see this place . . . commit it to memory. Think of every detail. Remember that in the future when you begin to feel anxious, you can just close your eyes, take a deep breath, and allow yourself to smile and remember this anxiety-free happy place that you have created for yourself.

As you hold on to the feeling of peace and joy in this place, take a deep breath in . . . and as you exhale, listen to the sound of your breath . . . continuing to breathe in and out. Remind yourself that you are safe in this moment and in your special place. Anxiety does not live here any longer.

(*Pause*)

When you are ready, start wiggling your toes and fingers, slowly open your eyes, smile, and stay with your breath.

Check in with yourself . . . how are you feeling?

Think of this happy place anytime you begin to experience anxiety and want to experience full relaxation.

BETWEEN-SESSION WORKSHEET

Daily Practice: Grounding Meditations

Use this worksheet to record your practice of grounding meditations throughout the week—the recommended practice is three to five times a week. Rate your anxiety before and after each practice, and then describe how you feel afterward.

Meditation	Date/Time	Anxiety Rating (1–10)		I feel . . .
		Before	After	

BETWEEN-SESSION WORKSHEET

Grounding Action Steps

This list of action steps can help regulate your nervous system and ground you when you're feeling panicked or anxious. Choose any of these action steps when you're feeling anxious, and rate your level of anxiety before and after each exercise.

Anxiety Rating Before (1–10)	Grounding Action Step	Anxiety Rating After (1–10)
	List five things you're thankful for.	
	Count to 10 or say the alphabet very slowly.	
	Notice your body. What are you wearing? How does your shirt feel on your chest? Wiggle your toes or feel the object you are sitting on.	
	Dig your heels into the floor, literally "grounding" them! Notice the tension centered in your heels as you do this. Remind yourself that you are connected to the ground.	
	Describe the steps in performing an everyday activity. For example, you might describe how to make your bed, do the dishes, cook your favorite meal, or tie a knot.	
	Think of the names of your friends or family members. How many can you name? How old are they? Can you spell their names?	
	Read something around you. Read the letters aloud forward and backward.	
	Think of an object and "draw" it in your mind or in the air with your finger. Try drawing a piece of fruit, a car, a house, or an animal.	

CHAPTER 9

Journaling, Affirmations, and Mirror Work

Journaling

Writing is a powerful tool for helping distract and manage the anxious mind. One of the most healing writing exercises is journaling, which is a wonderful way for clients to tap into their emotional self while gaining insight into their irrational thinking patterns. The more that clients use their journal to rewrite their story, reframe their thinking patterns, and replace automatic negative thoughts, the more likely it is that their healing will progress.

This works because of the power of neuroplasticity, which is the "muscle-building" part of the brain that gives people the ability to change and adapt by forming new neural pathways. The brain is always evolving and never stops changing in response to learning. All it takes is repetition and practice to strengthen these pathways and form new habits. This means that if your client continuously journals about the way they want to think and live, they can actually change their neural pathways. It takes consistent repetition to change these pathways, but over time, it becomes automatic—people literally become what they think and do!

If your client feels like prompts would make the journaling process a little less daunting, let them know that they're not alone! Anxiety can sometimes lead clients to feel "stuck" inside their own heads, which makes it difficult to know where to begin when it comes to journaling. The following exercises contain helpful information that will get them into the practice, including many different journal prompts that can act as therapeutic guidelines and allow clients to better express themselves.

PSYCHOEDUCATIONAL HANDOUT

My Journaling Practice

Starting the practice of journaling may seem a little daunting, but there are many ways to make the activity easier—so you can get a start on improving your well-being! Here are some ways you can start the journaling process.

1. **Start by getting something to write on:** It doesn't have to be one of those fancy spiral-bound journals. You can take an office notepad and just start writing. Or better yet, go to the dollar store and buy one of those notebooks you used in grade school. It's important that you use paper only when journaling. Try to avoid screens and write in a notebook instead.

2. **Set a timer:** Give yourself a time limit of 5–10 minutes to write down your thoughts on paper. When you have a set time, you're more likely to focus, and you'll get more out of the time you spend writing.

3. **Schedule it:** If you have a busy calendar or you struggle to prioritize journaling (and you know it is a helpful coping tool for you), schedule "journaling time" into your daily routine. Remember, 5–10 minutes daily is all you need. Look at your daily schedule and see when you are more likely to have time to write and when it will be most helpful. Try it for at least 21 days and watch it become a daily healing habit.

4. **Add prompts if you need to:** If you feel like prompts would make the journaling process a little less daunting, you're not alone! Anxiety can sometimes make you feel "stuck" inside your own head, which makes it difficult to know where to begin when it comes to journaling. Ask your therapist if you need help getting started.

THERAPEUTIC WORKSHEET

Journal Practice

Write in the space below for five minutes. Remember, there are no rules when it comes to journaling—just get out whatever thoughts are going through your mind at the time. Try to keep going until you feel you have written what needs to be said. Remember not to judge what you're writing. Acknowledge that whatever is coming out of your mind is 100 percent okay and that these are just *thoughts* that do not have any power over you. This can help you learn how to let go of situations that bother you and free your mind from distracting thoughts. It's up to you whether or not you would like to share what you write with your therapist.

_____

_____

_____

_____

_____

_____

_____

_____

_____

_____

_____

_____

_____

_____

_____

_____

_____

_____

BETWEEN-SESSION WORKSHEET

Self-Reflective Journal Prompts

These journal prompts are self-reflective in nature and will help you gain more insight into your anxious thought patterns, limiting beliefs, and fears. Circle the prompt of your choice and write your response for 5–10 minutes. Choose one journal prompt daily. It's up to you whether or not you would like to share what you write with your therapist.

What do you feel anxious about right now? Are these thoughts 100 percent true? Why or why not?	List someone who makes you feel nervous or anxious when you're around them. What aspects of this person do you have the power to change?	Write about a time when you were feeling anxious and it went away. What did you do to help yourself?
How do you know that you're feeling stressed or anxious?	Imagine your best friend has the same anxiety symptoms that you do. What would you say to that friend to help support them through it?	What secrets are you keeping? Are these secrets affecting your mental health? Why or why not?
How could you see your anxiety as helpful?	Write about the last time you cried. What caused you to cry?	If you weren't afraid, what are 5 things you would do? What are some ways could you overcome this fear?
What kinds of situations create anxiety for you? What aspects of these situations do you have the power to change?	Write a letter to your inner child. What advice can you give this child to better navigate their mental health? How could you nurture this child?	Have you been in a situation in which you thought you wouldn't survive? How did you get through that difficult time?

BETWEEN-SESSION WORKSHEET

Healing Journal Prompts

These journal prompts are healing in nature and will help you change the false narrative you hold in your mind that may be contributing to your anxiety. Circle the prompt of your choice and write your response for 5–10 minutes. Choose one journal prompt daily. It's up to you whether or not you would like to share what you write with your therapist.

What is your happiest memory?	What are 10 things you are thankful for?	What was the funniest thing that happened to you recently?
What are some ways you could release negative energy from your life?	What would a perfect day look like for you?	Who are 3 people you could connect with right now if you needed help?
What makes you feel better after a bad day?	When was a time you were proud of yourself?	What is something positive you've learned about yourself recently?
What does your happy place look like?	If you could give your younger self advice, how could they better navigate their mental health?	What did you see today that was beautiful?
What are the 3 biggest obstacles you've overcome?	What made you smile today?	What was something that happened in your childhood that you are grateful for?
What are 3 things that have helped you when you felt stressed or anxious in the past?	What are the top 5 moments when you felt the happiest?	When do you feel the calmest?
What coping skills do you use to deal with anxiety?	What are 5 things you love to do?	Why are you glad to live where you live?

BETWEEN-SESSION WORKSHEET

Self-Esteem Journal Prompts

These journal prompts will remind you of how special you are and give a boost to your self-esteem, self-confidence, and self-worth. Circle the prompt of your choice and write your response for 5–10 minutes. Choose one journal prompt daily. It's up to you whether or not you would like to share what you write with your therapist.

What are some things that went well for you today?	What did someone help you with today?	What made you laugh today?
When was a time you felt you succeeded?	What is your favorite thing about yourself?	What are 3 things you appreciate about yourself?
What did you accomplish today?	What are 5 reasons that you are amazing?	What is something positive you witnessed today?
What made you feel peace today?	If your younger self could see you today, what would they be most proud of?	How did you have fun today?
What did you do for someone else today?	When was a time you felt good about yourself?	When do you feel most like yourself?
What is something you love about yourself?	When was a time you felt really excited?	What is something you do really well?
When was a time you felt very confident?	When was a time that you were a really good friend?	What are some ways you are thankful for your body?
What are your best character traits?	When was a time you overcame a big challenge?	What makes you unique?
Who are 10 people who love you?	What is something you did that made you proud?	What was a time you felt joy?

BETWEEN-SESSION WORKSHEET

Positive Mindset Daily Journal

These journal prompts will help you develop a more positive mindset and reframe your negative thoughts, providing you with a sense of control and empowerment. Throughout the next week, answer these questions daily to focus more on the aspects of your life that give you a sense of positivity, gratitude, and appreciation. Review your answers with your therapist.

I felt loved today when _____
_____
_____

Something good that happened to me today was _____
_____
_____

I felt grateful today because _____
_____
_____

A compliment that I would give myself today is _____
_____
_____

Two positive feelings that I experienced today were _____
_____
_____

I made someone else feel good when I _____
_____
_____

Something positive that someone said about me was _____
_____
_____

I had a negative thought about myself when _____
_____
_____

A different thought that I can have next time is _____
_____
_____

Something I can do to make tomorrow a better day is _____
_____
_____

Affirmations

Affirmations are short, powerful statements that provide clients with words of encouragement, help them challenge unhelpful or irrational thoughts, and offer them the motivation to overcome self-doubt. It is important to remember that what we think, we believe. Unfortunately, many anxious clients speak to themselves in a negative or critical way, which limits their ability to create the very things they want in life. Many of these irrational thought patterns stem from beliefs they learned in childhood, and to break these ingrained patterns, clients must practice retraining their brain to think and speak in a positive way. This is where the power of affirmations can help.

There are many ways that clients can use positive affirmations. Some people prefer to repeat a statement aloud, while others may prefer writing out or recording their affirmations. Clients can also choose a preferred time when they'd like to practice affirmations each day, but repeating them in the morning can start the day off in a positive way. Your clients should make this practice a regular habit, repeating the affirmations regularly. It may help to instruct them to create an alert on their smartphone so they will be notified of their favorite affirmation. Encourage them to write out the affirmation in the alert so they can be visually reminded throughout their day. The following resources contain several affirmations that you can review with your client to find what speaks to them.

PSYCHOEDUCATIONAL HANDOUT

How to Create Affirmations

Affirmations are short, powerful statements that can provide you with words of encouragement, help you challenge your unhelpful or irrational thoughts, and offer you the motivation to overcome self-doubt. Here are some tips on how you can make affirmations successful.

1. **Write your affirmations in the present tense.** Write affirmations about your life as if you already have what you want. This helps your mind visualize the outcome. For example, you can say, "I have an abundant life filled with love and joy" instead of "I will find love from others and create joy."

2. **Keep it simple.** You are more likely to remember a self-talk statement that is short and sweet.

3. **Make sure to state them as a fact, not a possibility.** Consider "I am full of wealth" rather than "I can make money."

4. **No negative words, only positive!** Do not use negatives in your affirmation. Instead of "I will not say mean words to myself," try "I accept myself for all that I am and that is worthy."

5. **Make it mean something.** Make sure the affirmations you choose speak to you and are meaningful on all levels.

6. **Be persistent and creative.** Try to work with only three to five positive affirmations at a time. Use these for a couple of weeks before switching to new ones.

THERAPEUTIC WORKSHEET

My Affirmation List

Use this worksheet to start creating a list of affirmations that resonate with you. Next to each affirmation that you create, rate how much you believe it on a scale of 1 ("not at all") to 100 ("totally believe it"). For any affirmations that you rate at a 50 or lower, discard it or reword it so it becomes more believable. Here are some examples of affirmations, but the possibilities are endless, so be sure to create statements that speak to you.

- I am dealing with this the best way I can.
- I let things go.
- I am safe, I am strong, and I am well.
- I am capable of handling anything.
- I am enough.
- I love myself yesterday, today, and tomorrow.
- I am worthy of love.
- Productivity does not define my value.
- I forgive myself.

Affirmation	Believability (0–100)

Copyright © 2024 Alison Seponara, *The Anxiety Healer's Guide for Clinicians*. All rights reserved.

THERAPEUTIC WORKSHEET

My Strengths

When you feel sad, anxious, or scared, it's easy to start having negative thoughts about yourself. In these moments, positive affirmations can remind you of your strengths and change negative thinking patterns. To get started, use the following list of strengths and circle the ones you would describe yourself with.

Adventurous	Loyal
Artistic	Motivated
Brave	Nice
Caring	Open-minded
Cool	Organized
Curious	Patient
Determined	Persistent
Disciplined	Reliable
Easy-going	Respectful
Empathetic	Responsible
Flexible	Selfless
Friendly	Smart
Funny	Sociable
Generous	Talented
Grateful	Thoughtful
Hardworking	Trustworthy
Helpful	Understanding
Honest	Unique
Independent	Warm
Intelligent	Wise
Kind	Witty
Leader	Other: _____

BETWEEN-SESSION WORKSHEET

List of Affirmations

Using the strengths you've discovered in the *My Strengths* worksheet, come up with at least five affirmations you can use next time you're feeling sad, anxious, scared, discouraged, or hopeless. Hang this list somewhere in your home where you can see it, and practice saying these statements to yourself in the mirror every day.

1. _____
   _____

2. _____
   _____

3. _____
   _____

4. _____
   _____

5. _____
   _____

BETWEEN-SESSION WORKSHEET

Five Things I Love About Myself

Write five things that you love about yourself in the spaces below. Try to think of at least three *internal* qualities you love about yourself—for example, maybe it's your sense of humor, your intelligence, or your big heart.

1. _____
   _____

2. _____
   _____

3. _____
   _____

4. _____
   _____

5. _____
   _____

PSYCHOEDUCATIONAL HANDOUT

Daily Affirmations Practice

By repeating affirmations to yourself, you are training your subconscious mind to attract the things you want in life. Change doesn't happen overnight, so it's important to stay consistent and be patient. With regular practice, you'll begin to notice a shift over time. Here are some ways to make your affirmation practice most effective.

1. Make short sentence cards for your affirmations based on your current need or situation.

2. Take the cards with you and read them out loud several times a day.

3. Use sticky notes or notecards to post affirmations on places you frequent, such as your desk, computer screen, bathroom mirror, or car's dashboard.

4. Record a voice memo on your phone that you can listen to whenever you want.

5. Write the same positive statement a set number of times.

6. Repeat your affirmations aloud in the mirror, shower, or car.

BETWEEN-SESSION WORKSHEET

Affirmation Log

Use this worksheet to record your daily affirmations. Schedule specific times each day to practice—the recommended practice is two to three times a day for just a few minutes. You may add or subtract time as needed. Use the log to track when you complete your daily affirmation check-in and note which affirmations you choose. At the end of each week, describe how you feel.

	Monday	Tuesday	Wednesday	Thursday	Friday	Saturday	Sunday
Week 1							

Affirmations used:

I feel . . .

	Monday	Tuesday	Wednesday	Thursday	Friday	Saturday	Sunday
Week 2							

Affirmations used:

I feel . . .

	Monday	Tuesday	Wednesday	Thursday	Friday	Saturday	Sunday
Week 3							

Affirmations used:

I feel . . .

	Monday	Tuesday	Wednesday	Thursday	Friday	Saturday	Sunday
Week 4							

Affirmations used:

I feel . . .

Mirror Work

There is no hiding from ourselves when we're in front of the mirror. As Louise Hay (2016) states, "The mirror reflects back to us the feelings we have about ourselves" (p. 2). When we stand in front of a mirror, we are confronted with a unique opportunity to examine the relationship we have with ourselves. It is in this intimate space that we catch a glimpse of our inner critic—that voice inside our head that judges and criticizes us. Mirror work can initially make us feel uneasy because it shines a light on this inner critic, exposing it for all to see. However, this discomfort can also serve as a catalyst for growth and healing.

To begin shifting how clients see themselves, you can introduce them to the practice of mirror work, in which they simply look in the mirror and repeat words of self-love and encouragement to themselves. By encouraging gentle and loving self-talk, clients can become more in touch with their inner self and develop a more compassionate and forgiving connection with themselves. Mirror work may feel awkward or uncomfortable at first, but it is one of the most effective healing methods for the inner critic. By using mirror work in session, your clients will begin to peel back any limiting beliefs rooted in fear and insecurity.

IN-SESSION PRACTICE

Introducing Mirror Work

Introduce this exercise to your client in session. With practice, they will be able to add it to their healing toolkit to help reduce their anxiety symptoms. Encourage your client to rate their anxiety before and after the exercise.

Clinician Script

Mirror work is a powerful tool for healing anxiety. It involves looking at yourself in a mirror and speaking positive affirmations aloud. Mirror work helps you to connect with yourself on a deeper level and develop self-acceptance and self-love. For this practice, I'll give you a small mirror and ask that you look at yourself for a few minutes. Ready? Let's begin.

- - - - -

Look yourself in the eye and don't look away.

(*Pause*)

What type of emotions are coming up for you as you do this?

(*Pause*)

You may feel awkward, unsettled, embarrassed, or emotional. It's okay to feel emotional—let yourself feel whatever comes up.

You may even feel some critical thoughts arise. Why? Because that inner critic is coming to the surface and telling you everything that it believes is "wrong" with you.

(*Pause*)

Try to silence that inner critic, repeating these affirmations after me as you continue gazing into the mirror:

"I am worthy of love and belonging."
"I am safe."
"I am enough."
"I forgive myself."
"I deserve love and happiness."

(*Pause*)

Although this may feel awkward or uncomfortable at first, stick with it! With time, this will help silence that inner critic and allow you to develop a more nurturing relationship with yourself.

THERAPEUTIC WORKSHEET

Mirror Work Reflection

Write down any thoughts or emotions that came up during the mirror work exercise in the space provided. How did you feel? Was it uncomfortable? Why or why not?

_____
_____
_____
_____
_____
_____
_____
_____
_____
_____
_____
_____
_____
_____
_____
_____
_____
_____
_____
_____
_____
_____

BETWEEN-SESSION WORKSHEET

Mirror Work

Mirror work brings your inner critic to the surface. By learning more about this inner critic, you can begin to shift your negative thinking patterns and use affirming self-statements to kick that critic to the curb. Over the next week, try to dedicate at least five minutes a day to the following mirror work practice. Make sure to do this exercise in private so you are not disturbed. (Lack of privacy may make it harder to be 100 percent vulnerable.) At the end of each mirror work exercise, write down any thoughts or emotions that came up for you and review them with your therapist.

1. Refer to your completed *List of Affirmations* worksheet before getting started.

2. Sit or stand in front of a mirror for five minutes. Setting a timer may be helpful.

3. Look yourself in the eyes and don't look away.

4. As you look into the mirror, note what emotions come up for you. You may feel awkward, unsettled, embarrassed, or self-critical. Remember, that's because mirror work exposes your inner critic.

5. To reestablish a connection with the loving, compassionate part of yourself instead, pick any of the statements on your *List of Affirmations* and repeat them out loud as you continue gazing into the mirror.

6. Repeat these statements slowly, taking a moment in between each one. What do you notice coming up for you? What feelings arise? Does that inner critic try to overshadow these loving statements?

7. After five minutes is up, take out your journal or any piece of paper and write down any thoughts or emotions that came up for you.

CHAPTER 10

Visualization Strategies

Visualization is a powerful tool that uses mental imagery to help clients relieve stress and achieve a more relaxed state of mind. When clients experience panic, their thoughts wander, and their mind clings to worst-case scenarios, jumps to conclusions, and focuses only on the negative. This only adds to their sense of fear. By practicing visualization techniques, you can help clients focus their attention on calming and serene images, expanding their ability to rest and relax.

One particular visualization technique included in this chapter is creative visualization, which can help clients promote success in every area of life. Shakti Gawain (2010), a pioneer in the field of personal development and an internationally renowned teacher of consciousness, describes creative visualization as "the technique of using your imagination to create what you want in your life" (p. 3). Gawain suggests that everything is energy, including thoughts. This means that people attract into their lives what they think about the most, believe in the strongest, and imagine the most vividly. By using creative visualization, clients will be able to create a clearer image of the goals they want to manifest—emotionally, mentally, physically, and spiritually. As they begin to visualize the life they want, they can more easily reach their goals and achieve the outcomes they desire.

Another visualization strategy included in this section is a series of guided imagery meditations. Guided imagery serves as a form of distraction that helps clients redirect their attention away from what may be stressing them at the moment. It is essentially a nonverbal suggestion provided to their body and subconscious mind that a peaceful, safe, and relaxing environment they envision in their mind is real. These scenes then become a learned cue when they are beginning to feel anxiety creep in, prompting them to recall memories and sensations from past guided imagery practices. The visualization strategies included in this chapter are:

- Creative visualization
- Future self visualization

- Guided imagery meditation
    - The clouds
    - The pond
    - The white sandy beach
    - The safe place
    - The happy memory

THERAPEUTIC WORKSHEET

Creative Visualization

Creative visualization is a technique in which you imagine creating what you want in life. As you begin to visualize the life you want, you may even begin to experience the emotions connected to this image—as if it were true. Take a few moments to visualize a future goal. Use the space below to describe exactly what you envision, using as much detail as possible.

1. **Choose a goal:** Think of something you would like to have, work toward, or create. Some examples might be a job, a relationship, a house, a happier state of mind, improved health, a better physical condition, or a change in yourself. It really can be anything that you would like to achieve!

2. **Create a clear mental picture of the outcome exactly as you want it:** Make this image vivid, with lots of details. Think of it in the present tense, as if it already exists the way you want it to be. It's almost as if you are already living it. How do you desire it to be right now? Make yourself a part of the picture and see yourself enjoying the outcome.

3. **Give the image positive energy:** Think of this image in an encouraging way. Use strong, positive affirmations to imagine yourself receiving or achieving this goal. Focus on this goal and imagine that the best possible outcome is happening right now. Try to remove any doubts or disbelief you may have, at least for the moment. Practice believing that what you desire is very real and possible.

_____

_____

_____

_____

_____

_____

_____

_____

_____

BETWEEN-SESSION WORKSHEET

Daily Visualization

Focus on the image you created in the previous *Creative Visualization* worksheet and review your description of it throughout your week. Without putting pressure on yourself to make something happen, just bring the story along with you as you move throughout your day. Reread your visualization once in the morning and once in the evening, using this log to keep track of when you have completed the exercise. At the end of each day, describe how you feel.

Day	Time		I feel . . .
	a.m	p.m.	
Monday			
Tuesday			
Wednesday			
Thursday			
Friday			
Saturday			
Sunday			

THERAPEUTIC WORKSHEET

Future Self Visualization

Visualizing your best future self will give you the motivation and inspiration you need to stay focused on your goals whenever you feel stuck or anxious. Use this worksheet to keep yourself accountable to becoming this person.

Imagine you are meeting your future self for the very first time. What does this person look like? What are you wearing?

_____

_____

_____

Where does your future self live? What does this place look like? Is anyone else there?

_____

_____

_____

What is your future self's life like? What do you love about your life?

_____

_____

_____

What do you need to do to get from where you are now to where your future self is?

_____

_____

_____

What would be most helpful to move toward these future goals? Listen to what your future self has to tell you.

_____

_____

_____

Think of any anxieties or stressors you are experiencing in your *current* life. Let your future self know that you are feeling worried, scared, and hurt about these situations. Ask your future self, "How did you overcome these challenges?"

_____

_____

_____

Finally, ask your future self to tell you one word that is important for you to remember when you're feeling down or when you need support to keep going toward your goals. What is that one word?

_____

BETWEEN-SESSION WORKSHEET

The Vision Board

A vision board, or dream board, is a board or display filled with images, pictures, and affirmations of your goals, dreams, and desires—how you want to feel, what you want to do, what you want to have, and who you want to be. Many people believe that they can attract into their life what they envision. Try using a vision board to manifest your ideal life on paper. This worksheet will walk you through the steps needed to create your own vision board.

Materials

- Poster board (or a notebook, journal, box, or envelope)
- Scissors
- Glue
- Coloring utensils
- Magazines, newspapers, or catalogs
- Crafting supplies such as sequins, glitter, or pom-poms (*optional*)

Instructions

1. Look through old magazines, newspapers, or catalogs and use scissors to cut out any photos, quotes, symbols, or words that resonate with you and your dreams, goals, or intentions in life.

2. When you feel like you have enough images, lay them out on the poster board and decide what looks most visually appealing to you. There is no right or wrong way to do it!

3. Once you've decided on a layout, glue the images onto the poster board. Use any additional crafting materials to decorate the board and write any additional words of encouragement.

4. Hang your vision board in a place where you will see it often. Since the board displays everything you want to focus on in life, it is helpful to hang it up in a place where you will see it every day. The more you look at your vision board, the more you will stay focused on achieving your goals and dreams. You are bringing energy to those goals to make them a reality!

Variation

If you don't want to gather all the materials for a physical vision board, you can use online platforms like Pinterest or Canva to create a digital collage of your goals. You can store this collage on your phone or print out your digital vision board and put it on your wall, mirror, or refrigerator.

IN-SESSION PRACTICE

The Clouds

Introduce this guided imagery visualization to your client in session. With practice, they will be able to add it to their healing toolkit to help reduce their anxiety symptoms. Encourage your client to rate their anxiety before and after the exercise.

Clinician Script

This guided imagery visualization will help put you in a more relaxed state of mind and expand your ability to focus your attention on calming and serene images rather than intrusive anxious thoughts. Now, let's begin.

- - - - -

Begin by getting into a comfortable position and closing your eyes. As you start to relax, create a picture in your mind. It is a warm summer day. Imagine that you are laying on a blanket outside. The blanket is soft, and the grass feels like a bed of clouds underneath your body.

(*Pause*)

You look around and see trees beside you—a mix of leafy trees, full of different shades of green. Between the leaves, you see large tree trunks stretching into the sky, and you watch as the branches move up and down with the wind. The breeze feels cool and comfortable.

(*Pause*)

You look up and notice the bright blue sky above. You look at the clouds, noticing their different shapes. Some are round, fluffy clouds. Others are long, thin, and wispy clouds. Some clouds look as though they were drawn with a paintbrush across the sky. The clouds drift by slowly, smoothly, silently. As the clouds float by, the warm sun shines down and relaxes you, creating a calm, sleepy feeling.

(*Pause*)

As you sink into your cozy blanket on the soft grass, you begin to feel your body relaxing bit by bit . . . feeling your muscles release . . . letting go of any tension . . . breathing in the healing air . . . breathing out any worries.

(*Pause*)

Take a deep breath in, and as you exhale, allow your body to relax. Continue to breathe slowly . . . deeply.

(*Pause*)

You close your eyes and listen to the calming sounds all around you. You hear the sound of birds singing in the distance . . . the wind in the trees . . . and the faint sound of children playing and laughing.

(*Pause*)

Imagine stillness in your body . . . and stillness in your mind. In this moment, you are safe. In this moment, you are at peace. There is nothing else to do but enjoy this moment, gazing up at the sky, watching the clouds drift by, and enjoying this beautiful day.

(*Pause*)

Take another deep breath in through your nose and slowly exhale for 8 . . . 7 . . . 6 . . . 5 . . . 4 . . . 3 . . . 2 . . . 1 . . . allowing your body to completely relax.

(*Pause*)

When you are ready to leave this peaceful place, slowly begin to return your awareness to the present. As you breathe, allow your body to reawaken. Bring yourself back to your usual level of awareness, wiggling your toes and fingers and feeling the energy flowing through your muscles. While returning to a state of alertness, keep with you the feeling of calm and relaxation you have just experienced. When you are ready, open your eyes and return to your day feeling refreshed.

IN-SESSION PRACTICE

The Pond

Introduce this guided imagery visualization with your client in session. With practice, they will be able to add it to their healing toolkit to help reduce their anxiety symptoms. Encourage your client to rate their anxiety before and after the exercise.

Clinician Script

Visualization is a powerful tool that allows you to create mental images that evoke feelings of peace and tranquility. This guided imagery will help put you in a more relaxed state of mind and expand your ability to focus your attention on calming images of a meadow and serene pond. Now, let's begin.

Begin by getting into a comfortable position and closing your eyes. Take a few deep breaths, noticing how your body feels. Take another deep breath in, hold it, and then breathe out, releasing tension. As you visualize the following scene, let your body and mind become more and more relaxed with each moment.

Imagine yourself walking outdoors. It is not too warm and not too cold . . . it's a perfect spring day. You are walking through a grove of trees, their leaves moving in a slight breeze. You see a pond straight ahead glistening in the sun. You notice a wooden bench overlooking the private and serene woodland pond. As you walk toward the bench, you notice the fish jumping up from the water, the butterflies fluttering around the trees, and the dragonflies flying back and forth as they soar over the water.

(*Pause*)

As you take a seat on the worn wooden bench, you close your eyes and take a deep breath in . . . exhaling slowly . . . allowing your body to completely relax. Listening to all that surrounds you, you notice the frogs singing on their lily pads and the busy bees buzzing around in the spring air. The birds begin chirping as you look up to the sky. You feel the warmth of the sun on your face. And there you are . . . totally still . . . utterly at rest.

(*Pause*)

Continue to breathe slowly and deeply. Breathe in again, and as you exhale, allow your body to relax. Smell the grass . . . the wildflowers . . . and the smell of the sun on the earth.

(*Pause*)

You notice a small ladybug climbing up a blade of grass, pausing for a moment, and then flying away. You remember that ladybugs are good luck, and a small smile forms on your face.

(*Pause*)

You look around again to see the sights around you. You notice the pond water rippling as the fish swim back and forth. You see the blue sky above you . . . the clouds slowly drifting by.

(*Pause*)

As you look out toward the lush green meadow, you notice a deer peeking out through the trees as it grazes in the distance. The deer raises its head to look at you, sniffing the breeze, and then turns, disappearing silently into the trees.

(*Pause*)

Now it is time to leave the pond and return to the present. Wiggle your toes and fingers and feel the surface beneath you. Listen for the sounds around you. Open your eyes to look around. Take a moment to stretch your muscles and allow your body to reawaken. When you are ready, return to your usual activities, keeping with you a feeling of peace and calm.

IN-SESSION PRACTICE

The White Sandy Beach

Introduce this guided imagery visualization with your client in session. With practice, they will be able to add it to their healing toolkit to help reduce their anxiety symptoms. Encourage your client to rate their anxiety before and after the exercise.

Clinician Script

This white sandy beach visualization will help you expand your ability to focus your attention on more calming images. The more you engage with this serene mental landscape, the more easily you will be able to access a state of calm and relaxation. Now, let's get started.

- - - -

Begin by getting into a comfortable position and closing your eyes. Relax your face and release any tension in your forehead, neck, and throat. Allow your breath to slow down, taking a deep breath in . . . and a long breath out.

(*Pause*)

Allow your entire body to rest heavily on the surface where you are sitting. Now that your body is fully relaxed, let's take a trip to your favorite beach. Imagine you are walking through a beautiful, tropical forest toward a white sandy beach. You feel safe, calm, and relaxed.

(*Pause*)

You hear the waves up ahead, and you can smell the ocean. As you make your way through the trees, you finally see the blue ocean water, and soft waves lap at the shore as the tide gently rolls in. The air is warm and it flows through your body. You feel a pleasant, cool breeze blowing through the trees.

(*Pause*)

You come out of the forest into a long stretch of white sand. The beach is wide and long, and as you take off your shoes, you notice that the sand feels like soft powder on your feet. You smell the clean, salty air and see the incredible aqua color of the ocean ahead.

(*Pause*)

As you approach the water, you can feel the mist from the ocean on your skin. You walk closer to the waves and feel the sand becoming wet and firm. Just then, a wave crashes and the water comes quickly to shore. You step forward, feeling the cool water provide you relief from the warm sand before it returns back into the abyss of the ocean.

(*Pause*)

You walk a bit farther into the clear blue water . . . just enjoying the ocean for a few minutes . . . noticing its pleasant, relaxing temperature . . . becoming more and more relaxed. The water provides welcome relief from the hot sun . . . cool but not cold.

(*Pause*)

As you begin to walk along the edge of the water, you are free of worries, free of stress, and full of peace.

(*Pause*)

When you are ready, walk out of the water toward a comfortable lounge chair and towel, just for you. Feel the weight of your body sinking into your beach chair, the warmth of the sand on your feet, and the large umbrella keeping you slightly shaded, creating the perfect temperature.

The only thing to do right at this moment is to be still and enjoy the sun on your face, the breeze in your hair, and the waves on your toes.

(*Pause*)

Allow all of your stress to melt away . . . feeling calm . . . feeling at peace . . . feeling refreshed.

(*Pause*)

When you are ready to leave the beach, do so very slowly as you bring yourself back to your usual level of alertness and awareness. Open your eyes, wiggle your toes and fingers, and feel the surface beneath you. Listen for the sounds around you as you become fully alert. Take a moment to stretch your muscles and allow your body to reawaken. When you are ready, return to your usual activities, keeping with you a feeling of peace and calm.

IN-SESSION PRACTICE

The Safe Place

Introduce this guided imagery visualization with your client in session. With practice, they will be able to add it to their healing toolkit to help reduce their anxiety symptoms. Encourage your client to rate their anxiety before and after the exercise.

Clinician Script

This safe place guided imagery meditation will help alleviate anxious feelings and bring a sense of calm and balance to your mind and body. Your safe place can be anywhere you feel comfortable and secure. By fully immersing yourself in this mental sanctuary, you can transport yourself to a place of tranquility and peace. Now, let's get started.

- - - -

Begin by getting into a comfortable position and closing your eyes. For the next few moments, focus on calming your mind by paying attention to your breath. Allow your breath to center and relax you. Relax your face and release any tension in your forehead, neck, and throat. As your breath begins to slow down, focus on your belly, feeling it rise up on the inhale and fall down on the exhale. Take a deep breath in through your nose . . . and a long breath out through your mouth.

(*Pause*)

Begin to create a picture in your mind of a place where you feel safe and completely at ease. Where is this place? Maybe it is somewhere outdoors or indoors. Maybe it is a place you have been before or somewhere that you long to go.

(*Pause*)

Now picture some more details about your safe place. What does this place look like? Is it small or large? What colors, shapes, or objects do you see? Is there water? Are there plants? Animals? What are the beautiful things that make your place enjoyable?

(*Pause*)

Are you alone in your safe place, or are there others with you? Notice that in this place, whether you are alone or with company, you feel completely safe.

(*Pause*)

Focus on the relaxing sounds around you in this peaceful place. What sounds do you hear? Perhaps it is silent. Pay attention to the sounds that are more noticeable and those that are more subtle. Are these sounds far away or close by?

(*Pause*)

Drifting away into this safe place, you feel more and more relaxed . . . more and more at peace.

(*Pause*)

Next, focus on what sensations you can feel in this place. Do you notice the feeling of the earth beneath you? Can you feel the texture of whatever you are sitting or lying on? What is the temperature like here? Can you feel a breeze, or does it feel still?

(*Pause*)

Think about any smells or tastes you may notice about your place. Maybe you notice the scent of lavender, peppermint, lemon, or rosemary in the air. Perhaps you can taste freshly baked cookies. Whatever you smell or taste, take a moment to savor it.

(*Pause*)

Maybe you would like to give your peaceful and safe place a name. It might be only one word or it could be a phrase that you can use to bring that image back anytime you need to.

(*Pause*)

Now that you have a picture of this safe place, imagine yourself there. What are you doing in this calming place? Maybe you are just sitting and enjoying the peaceful feeling of this moment. Maybe you are walking around or doing something you enjoy. Whatever it is, picture yourself in this place, completely at peace, becoming more aware of your breath as you slowly breathe in . . . and slowly breathe out.

(*Pause*)

This is a safe place . . . a place of calm . . . a place of peace . . . a place where you have no worries, cares, or concerns. A place where you can simply enjoy just being, with no pressures, no deadlines, no pain, no fears . . . only love and safety.

(*Pause*)

Rest in your safe place for a while, enjoying the peacefulness and serenity, and when you are ready to leave, slowly begin to turn your attention back to the present moment. As you bring yourself back to your usual level of alertness and awareness, begin to open your eyes, wiggle your toes and fingers, and feel the surface beneath you. Listen for the sounds around you as you become fully alert. As you return to this moment, remember to keep with you the feeling of calm from your safe place. File away this imaginary place in your mind . . . it will always be there for you the next time you need it.

IN-SESSION PRACTICE

The Happy Memory

Introduce this guided imagery visualization with your client in session. With practice, they will be able to add it to their healing toolkit to help reduce their anxiety symptoms. Encourage your client to rate their anxiety before and after the exercise.

Clinician Script

When you are consumed by anxiety, your thoughts tend to spiral and become overwhelming. However, by intentionally redirecting your attention toward a positive memory and immersing yourself in it, you can shift your focus and create a sense of calmness and inner peace, helping rewire your brain over time.

- - - -

Begin by getting into a comfortable position and closing your eyes. Take a deep breath in through your nose . . . and out through your mouth. Now think back to a time when you felt utterly happy and carefree. Visualize every detail of that memory. If you don't remember a specific detail, fill it in with whatever comes to mind.

(*Pause*)

Where were you?

What were you doing?

What were you wearing?

Who were you with?

What was the environment like?

(*Pause*)

Picture the location of the memory and everything about it, making special note of what you see, hear, touch, smell, and taste from this memory.

(*Pause*)

As you notice more and more details about this happy time in your life, feel yourself become more and more relaxed.

(*Pause*)

Once the image is complete in your mind, spend a few minutes just enjoying the memory.

(*Pause*)

When you are ready, slowly open your eyes and come back to this moment. Wiggle your toes and fingers and feel the surface beneath you. Listen for the sounds around you as you become fully alert. Take a moment to stretch your muscles and allow your body to reawaken. You are ready to face the rest of your day with a fresh mind.

BETWEEN-SESSION WORKSHEET

Daily Practice: Guided Imagery Meditation

Use this worksheet to record your practice of guided imagery meditation throughout the week—the recommended practice is at least three times a week. Put on some relaxing music (instrumental or meditative is best) and find a quiet place with no distractions. Rate your anxiety before and after each practice, and describe how you feel afterward.

Meditation	Date/Time	Anxiety Rating (1–10)		I feel...
		Before	After	

CHAPTER 11

Supplemental Holistic Healing Remedies

In this book, we have introduced and practiced cognitive restructuring, breathwork, grounding exercises, journaling, affirmations, mirror work, and visualization strategies that clients can include as part of their toolkit. Now, let's take a look at some other supplemental healing tools that clients might find effective.

Portable Items

- Activity book (e.g., connect-the-dots, mazes, word searches, I-Spy)
- Adult coloring book
- Art supplies (e.g., paper and crayons, markers, paint)
- Book of inspirational sayings
- Bubble wrap
- Calming or soothing music
- Eye mask
- Fidget toy
- Heating pad
- Hot tea (especially anxiety healing flavors like chamomile, peppermint, valerian root, lemon balm, lavender, and passion flower)
- Journal or notebook
- Noise-canceling headphones
- Silly putty
- Small bag of sand
- Stress ball
- Stuffed animal
- Water bottle
- Weighted blanket

Aromatherapy*

- Bergamot
- Chamomile
- Clary sage
- Frankincense
- Grapefruit
- Holy basil
- Jasmine
- Jatamansi
- Lavender
- Lemon
- Orange
- Patchouli
- Peppermint
- Rose
- Valerian
- Vetiver

Apps

- 1010!
- Breathe2Relax
- The Breathing App
- Breathwrk
- Buddhify
- Calm
- Dare
- Headspace
- Insight Timer
- The Mindfulness App
- Mindshift CBT
- Power of Calm
- Root'd
- Simple Habit
- The Tapping Solution
- Wim Hof Method

Videos

This will vary widely from client to client, but the idea is to identify online videos, movies, or TV shows that help the client feel calm when they're anxious. Here are some ideas:

- Progressive muscle relaxation videos
- Grounding videos (e.g., the five senses exercise)
- Breathwork videos (e.g., 4-7-8 breathing, alternate nostril breathing)
- Feel-good TV shows/movies (e.g., *The Office, Schitt's Creek, Friends, Bridesmaids*)

---

* It is important to remember that essential oils are very potent, so clients need to be cautious when using them. Not everyone will respond to essential oils in the same way.

Movement

- Exercise
- Hiking in nature
- Qigong
- Somatic movement (e.g., full-body stretching, hug yourself)
- Tai chi
- Swimming
- Walking
- Yoga

Books

- *The Anxiety and Phobia Workbook* by Edmund J. Bourne
- *The Anxiety and Worry Workbook: The Cognitive Behavioral Solution* by David A. Clark and Aaron T. Beck
- *The Anxiety Healer's Guide: Coping Strategies and Mindfulness Techniques to Calm the Mind and Body* by Alison Seponara
- *Be Calm: Proven Techniques to Stop Anxiety Now* by Jill P. Weber
- *Dare: The New Way to End Anxiety and Stop Panic Attacks* by Barry McDonagh
- *Dialectical Behavior Therapy Workbook: The 4 DBT Skills to Overcome Anxiety by Learning How to Manage Your Emotions* by David Lawson
- *Feeling Better: CBT Workbook for Teens: Essential Skills and Activities to Help You Manage Moods, Boost Self-Esteem, and Conquer Anxiety* by Rachel Hutt
- *The Highly Sensitive Person: How to Thrive When the World Overwhelms You* by Elaine N. Aron
- *How to Do the Work: Recognize Your Patterns, Heal from Your Past, and Create Your Self* by Nicole LePera
- *Love Yourself, Heal Your Life Workbook* by Louise Hay
- *The Power of Now: A Guide to Spiritual Enlightenment* by Eckhart Tolle
- *The Power of Vulnerability: Teachings of Authenticity, Connection, and Courage* by Brené Brown
- *The Relaxation and Stress Reduction Workbook* by Martha Davis, Elizabeth Robbins Eshelman, and Matthew McKay
- *Retrain Your Brain: Cognitive Behavioral Therapy in 7 Weeks: A Workbook for Managing Depression and Anxiety* by Seth J. Gillihan

- *Rewire Your Anxious Brain: How to Use the Neuroscience of Fear to End Anxiety, Panic, and Worry* by Catherine M. Pittman and Elizabeth M. Karle
- *When Panic Attacks: The New, Drug-Free Anxiety Therapy That Can Change Your Life* by David D. Burns
- *Worry Trick: How Your Brain Tricks You into Expecting the Worst and What You Can Do About It* by David A. Carbonell

Podcasts

- *The Anxiety Chicks* with Alison Seponara and Taylor Marae
- *The Anxiety Coaches Podcast* with Gina Ryan
- *Dear Therapists* with Lori Gottlieb and Guy Winch
- *Emotions Mentor®* with Rebecca Hintze
- *The Happiness Lab* with Laurie Santos
- *Help Me Be Me* with Sarah May Bates
- *Inside Mental Health* with Gabe Howard
- *Meditation Minis* with Chel Hamilton
- *On Purpose* with Jay Shetty
- *Oprah's Super Soul* with Oprah Winfrey
- *SelfHealers Soundboard* with Nicole LePera and Jenna Weakland
- *Therapy for Black Girls* with Joy Harden Bradford
- *This Changes Everything* with Sarah Rice
- *Unlocking Us* with Brené Brown
- *Your Anxiety Toolkit* with Kimberley Quinlan

Social Media

- Alison Seponara @theanxietyhealer
- Amanda E. White @therapyforwomen
- Annie Zimmerman @your_pocket_therapist
- Barb Schmidt @peaceful_barb
- Fight Through Mental Health @fightthroughmentalhealth

- Happiness Project @happinessproject
- Jeff Guenther @therapyjeff
- John Kim @theangrytherapist
- Julie Smith @drjulie
- Kimberley Quinlan @youranxietytoolkit
- Mel Robbins @melrobbins
- Nicole LePera @the.holistic.psychologist

Social Support

- Notecard with the name of at least one trusted support person and their contact information

*Supplements and Medications**

- Ashwagandha
- CBD oil
- Chamomile
- Kava
- L-theanine
- Magnesium
- Melatonin
- Omega-3 fatty acids
- Valerian root
- Vitamin B complex
- Vitamin D

Calm Sleep Environment

ITEMS

- Aromatherapy diffuser with calming essential oils
- Lavender spray on pillow
- Noise machine
- Sleep and wake-up light therapy lamp
- Soothing pillow
- Tea (herbal, decaf)
- Weighted blanket

---

* Please consult with a medical doctor before using any supplements or medications. The information provided here is not a replacement for medical advice. It is not meant to diagnose, treat, prevent, or cure any physical, mental, or emotional condition.

ACTION STEPS

- Avoid alcohol
- Avoid caffeine after 2:00 p.m.
- Avoid major physical activity and heavy meals close to bedtime
- Avoid napping too close to the evening
- Avoid watching TV, reading, or working in bed
- Journal about any worries before bed
- Keep regular sleep and wake times
- Keep the room a cool sleep temperature (approximately 65 degrees Fahrenheit)
- Meditate
- Reduce noise by wearing earplugs and silencing cell phone calls
- Take a bath before bed
- Use room darkening window treatments, heavy curtains, or an eye mask to eliminate as much natural light as possible

CHAPTER 12

The Client Healing Toolkit

Now it's time for your client to create their own anxiety healing toolkit. Remember that the goal of this workbook is to help your client create a *unique* toolkit that works specifically for their needs. To do so, have them use the anxiety rating scale from chapter 7 to keep track of the healing tools that consistently decrease their level of anxiety. If a technique doesn't work at first, encourage them to stick with it for a bit before moving on to another. It may take a lot of trial and error to establish the tools that work best, as there is no one-size-fits-all approach when it comes to healing anxiety. What works best for one person may not work best for the next person. Once your client has identified strategies to include in their toolkit, it is essential that they practice these strategies at the first sign of anxiety. They don't want to wait until their anxiety rating is at a 6, 7, or 8 because it will be much harder for them to come down from it. Once they get into the habit of using the tools at the first sign of anxiety, they are on the right path toward healing!

PSYCHOEDUCATIONAL HANDOUT

Sample Anxiety Healing Toolkit

Here is an example of what a complete healing toolkit may look like. Remember, everyone's anxiety looks different, so everyone's toolkit will look different!

Breathwork

- Square breathing
- Diaphragmatic (belly) breathing
- Alternate nostril breathing

Grounding Tools

- The five senses meditation
- Body scan meditation
- Progressive muscle relaxation
- Splashing cold water on my face
- Sucking on ice
- Yoga
- Painting with my niece and nephew
- Listening to soft music
- Organizing my room/belongings (e.g., putting away laundry, decluttering things)

Affirmations

- I am safe.
- I have felt like this before and have gotten through it.
- This feeling is only temporary.
- One step at a time.
- It is okay to just rest right now. There is nowhere I have to be.
- This will be over soon.
- Don't think; just breathe.
- My thoughts do not have control.

Journaling

- Self-reflective prompts two to three times/week
- Healing journal prompts two times/week

Mirror Work

- Morning affirmations once a day

Visualization Strategies

- The happy memory
- The white sandy beach

Portable Items

- Heating pad
- Eye mask
- Fidget spinner
- Silly putty
- Weighted blanket
- Journal

Aromatherapy

- Peppermint oil
- Citrus oils (e.g., lemon, orange, grapefruit)

Apps

- Insight Timer
- Calm
- 1010!
- The Tapping Solution
- Dare

Videos

- Progressive muscle relaxation videos
- Yoga for anxiety videos
- TV shows: *The Office, Schitt's Creek, Friends*
- Movie: *Bridesmaids*

Movement

- Yoga
- Go on a 10-minute walk once a day
- Work out while watching TV
- Walk on the beach
- Walking meditation

Books
- *The Anxiety Healer's Guide: Coping Strategies and Mindfulness Techniques to Calm the Mind and Body* by Alison Seponara
- *The Feeling Good Handbook* by David D. Burns
- *Love Yourself, Heal Your Life Workbook* by Louise Hay
- *The Anxiety and Worry Workbook: The Cognitive Behavioral Solution* by David A. Clark and Aaron T. Beck
- *How To Do the Work: Recognize Your Patterns, Heal from Your Past, and Create Your Self* by Nicole LePera
- *The Power of Vulnerability: Teachings of Authenticity, Connection, and Courage* by Brené Brown

Podcasts
- *The Anxiety Chicks* with Alison Seponara and Taylor Marae
- *Oprah's Super Soul* with Oprah Winfrey
- *SelfHealers Soundboard* with Nicole LePera and Jenna Weakland
- *Unlocking Us* with Brené Brown
- *Meditation Minis* with Chel Hamilton
- *On Purpose* with Jay Shetty

Social Media
- Kimberley Quinlan @youranxietytoolkit
- Julie Smith @drjulie
- Amanda E. White @therapyforwomen
- Alison Seponara @theanxietyhealer

Social Support
- Sister
- Mom
- Yoga community
- "Safe" friends
- Therapist
- Dog

Supplements and Medications
- Vitamin D
- B12
- CBD oil
- Ginger tea
- Magnesium

Calm Sleep Environment

- Noise machine
- Lavender spray on pillow
- Keep regular sleep and wake times
- Avoid coffee after 2:00 p.m.
- Meditate
- Take a bath before bed
- Use a darkening eye mask
- Keep temperature cool
- Avoid napping too close to the evening
- Journal about any worries before bed
- Avoid major physical activity and heavy meals close to bedtime

THERAPEUTIC WORKSHEET

Your Anxiety Healing Toolkit

Now that you have learned and practiced many different anxiety healing exercises, it's time to create your healing toolkit! Look back on all of your hard work and write down the action steps that work best in activating your relaxation response. What specific healing tools did you practice that helped reduce your anxiety levels? What made your nervous system feel more regulated? Remember to use the anxiety rating scale to assess how much distress you're feeling before and after each exercise. This can give you more insight as to which techniques may work best for you.

Breathwork

Grounding Tools

Affirmations

Journaling

Mirror Work

Visualization Strategies
Portable Items
Aromatherapy
Apps
Videos
Movement
Books

Podcasts
Social Media
Support System
Supplements and Medications
Calm Sleep Environment
Other

CHAPTER 13

The Therapist Healing Toolkit

Compassion Fatigue and Burnout

As therapists, we are committed to lending a compassionate ear, offering support, and actively listening to our clients' problems. But let's be honest—it can become emotionally draining and downright exhausting if we don't take proper care of our own mental and physical well-being. When this occurs, we can experience compassion fatigue, which refers to the emotional and psychological exhaustion that occurs when therapists consistently provide empathetic care to their clients. Compassion fatigue may even disrupt the therapeutic relationship and adversely affect client outcomes, so it is crucial to be aware of the signs and symptoms of compassion fatigue and take appropriate steps to prevent and address this issue. The following list includes many of the signs of compassion fatigue in therapists. Place a check mark next to any that you relate to.

- ☐ Chronic thoughts about clients, even outside the job
- ☐ Decreased ability to empathize with clients
- ☐ Increased irritability
- ☐ Feelings of cynicism or hopelessness
- ☐ General sense of emotional exhaustion
- ☐ Difficulty emotionally connecting with clients
- ☐ Sense of detachment
- ☐ Avoidance or isolation
- ☐ Nightmares, which may or may not be about work
- ☐ Inability to be fully present at work or with clients
- ☐ Substance misuse or abuse

If left unaddressed over time, compassion fatigue can lead to professional burnout, in which you experience chronic stress and exhaustion as a result of your work. Professional burnout is characterized by similar symptoms as compassion fatigue—fatigue, irritability, and a lack of motivation—but it can also negatively impact physical health, leading to problems such as headaches, stomach issues, and a weakened immune system. Burnout also increases the risk of anxiety, depression, and other mental health disorders (Risser, 2022).

Stressors that Contribute to Therapist Burnout

- Sociopolitical stressors
- Inability to detach from clients' issues and concerns
- Secondary trauma
- Exposure to aggressive, depressed, or suicidal clients
- Emotional fatigue
- Focusing on others' needs rather than one's own
- Being on-call
- Administrative task buildup
- Financial stress
- Client emergencies and crises
- Slow progress with certain clients

For therapists in particular, burnout can be especially detrimental. The following lists are all signs of more general burnout and burnout within the therapy profession in particular. Place a check mark next to any that you relate to.

SIGNS OF GENERAL BURNOUT

- ☐ Mental and physical exhaustion
- ☐ Workplace dread
- ☐ Irritability
- ☐ Struggles with time management
- ☐ Lack of sleep
- ☐ Performance decline

- ☐ Chronic anxiety
- ☐ Detachment
- ☐ Feelings of listlessness
- ☐ Low mood
- ☐ Physical ailments (e.g., headaches, muscle tension, gastrointestinal issues, chronic fatigue)
- ☐ Trouble concentrating
- ☐ Forgetfulness

SIGNS OF THERAPIST BURNOUT

- ☐ Dreading work
- ☐ Canceling appointments or showing up late
- ☐ Daydreaming or feeling distracted during appointments
- ☐ Feeling emotionally drained
- ☐ Feeling relieved when clients cancel
- ☐ Feeling overwhelmed
- ☐ Experiencing a decline in empathy
- ☐ Self-medicating or numbing out with alcohol or other behaviors (e.g., scrolling on social media, shopping)
- ☐ Feeling mentally distanced from one's job
- ☐ Feeling increased negativity, cynicism, or loss of purpose related to the job
- ☐ Experiencing a reduced professional efficacy
- ☐ Feeling anxious, worried, or depressed
- ☐ Struggling with sleep

Therapist Self-Awareness

As a therapist, it's essential to have a deep understanding of the various anxiety diagnoses, but it's equally important to recognize that anxiety is not limited to your clients. Mental health struggles are a universal human experience, and acknowledging your own battles with anxiety can actually enhance your ability to relate and empathize with your anxious clients. By becoming more aware of your own fears and worries, you can develop a heightened sensitivity to the complexities of anxiety. This self-awareness allows you to gain a firsthand perspective on how anxiety can impact individuals in different ways. For example, you may discover that certain situations or triggers elicit strong anxiety responses in you, while in other circumstances, you feel no anxiety at all. This insight can help you recognize that your clients have their unique anxieties as well. Just because something doesn't cause *you* anxiety doesn't mean it's not a genuine concern for them. Take time to reflect on the following questions to help you practice more mental health self-awareness.

1. What are some of your own fears and worries?
2. How do you cope with highly anxious moments?
3. What helps you feel regulated and grounded?
4. What situations or triggers elicit strong anxiety responses in you?
5. What helps you stay present with your clients as they experience helplessness, despair, uncertainty, disappointment, and loss?
6. How do you take care of yourself when a client becomes emotionally overwhelmed or shuts down?
7. What support systems do you have in place for times when you feel emotionally unsettled from your work?
8. In what ways have you been changed for the better by your work? Can you recall moments when you learned more about yourself as a result of your clinical work? Are there any specific clinical moments that have inspired you or provided you with a sense of hope?

Therapist Self-Care

As therapists, it is crucial for us to recognize the importance of making self-care a priority and having our own healing toolkit. Although we dedicate our lives to helping others, it is essential to remember that we also need to take care of ourselves in order to provide the best support for our clients. By prioritizing self-care and having our own healing toolkit, we not only enhance our own well-being but also strengthen our ability to create a safe and healing space for those we serve. This toolkit can include a variety of self-care practices, whether it's engaging in mindfulness and meditation, exercising, journaling, seeking therapy for ourselves, spending time in nature, or connecting with supportive communities. By actively practicing self-care, we replenish our own energy and prevent burnout, which can lead to more compassionate and effective therapeutic work. The following section includes self-care tips for the mind, body, and soul.

Self-Care for the Mind

- **Commit to a daily mindfulness practice:** Find meditation, breathwork, and grounding exercises that work for you.
- **Leave work at work:** When you leave work, try to disconnect. Make your home a space of calm and relaxation without the pressure to continue work. If you work from home, make sure to keep that area separate from the rest of the home.
- **Limit checking emails:** Check your emails during a specific time each day. Try to stick to that time and that time only.
- **Create work-life balance:** Invest in prioritizing time for self-care outside of work. Make sure this time is spent in a meaningful way where you can fully disconnect from work and relax. It is also important to establish a manageable work schedule; know what kind of clients may trigger you and take precautions beforehand.
- **Limit your caseload:** I know, I know. You want to help as many people as you can! But the truth is, having too high a caseload doesn't really help you or the client. You may actually be doing the client a disservice since you only have a certain amount of mental capacity before your focus is overridden with other client stories.
- **Take a mental health day:** Set a date to take time away from anything work related. Focus on other activities you enjoy, such as hiking, making art, or connecting with friends and family.
- **Work with your ideal clients:** Many therapists who work outside of their scope of expertise find their work exhausting and risk burnout much more frequently. It's

important to find a population of clients that are ideal to you, which will leave you feeling more fulfilled.

- **Learn a new skill:** Take a class, learn a new language, or try a new activity.
- **Seek out your own therapy:** Even the most well-balanced person needs help from time to time when it comes to coping with adversity in life. Therapists also experience these hardships and need the support to get through them.

Self-Care for the Body

- **Invest in healthy sleep hygiene habits:** This may look like getting at least seven hours of sleep per night; maintaining a consistent sleep and wake schedule; creating a relaxing bedtime ritual; keeping your bedroom cool, dark, and comfortable; removing electronic devices from the bedroom; avoiding foods that can disrupt sleep; limiting caffeine and alcohol; avoiding naps; and unplugging from electronics at least an hour before bed.
- **Limit alcohol:** If you need alcohol to get through the day, consider seeking some support.
- **Eat a balanced, healthy diet:** Increase your intake of whole foods, fruits, and vegetables. Limit sugar and processed foods.
- **Schedule breaks throughout your day:** During these breaks, fully disconnect from work for that time.
- **Move your body:** You don't need to go to the gym every day to experience the benefits of a little exercise. Any type of movement can help, whether it's taking long walks, doing yoga, dancing, playing with your kids, or even cleaning.

Self-Care for the Soul

- **Increase your social support:** Maintain a strong support system of friends and family members. Spend time with people who care about you and make you laugh. Make a standing appointment weekly or monthly with a friend, family member, or colleague.
- **Stay connected to your fellow therapists:** Join a case consultation group or commit to a monthly professional social activity. This community will help you to feel less isolated, and as a bonus, it can also help you expand your professional

capabilities. Join professional online forums like Facebook groups, LinkedIn groups, or *Psychology Today*. Connecting with others, especially those who understand the unique challenges of working in mental health, can help prevent isolation. Or even better, get together with fellow counselors for a counseling retreat or a peer supervision group.

- **Get mentorship and supervision:** Meeting monthly with a mentor, supervisor, or colleague can help you stay abreast of how you are doing in your career and reduce the risk of burnout. They may be able to help you delegate your responsibilities and reinforce the importance of self-care.

- **Connect with your community:** By connecting to something larger than yourself, you can get a sense of purpose and meaning in life. Get active in local recreation centers, yoga studios, religious entities, or volunteer organizations.

- **Set boundaries:** Understanding your limits and requirements is essential for prioritizing your own needs as a therapist. There are several areas that you must pay attention to:

    - **Boundaries with clients:** Make sure to have policies in place so clients know what to expect from therapy. This may include how to contact you between sessions and when this is appropriate, your cancellation policy and fees, the times when you will respond to calls and messages, and so forth. Make sure you communicate these boundaries with your clients—and more importantly, stick to them!

    - **Boundaries with work:** It is important to have solid boundaries between your work time and your personal time. If you work in an office, it can be helpful to notice when you cross a physical line on the way home from work, like a river or a highway, to set a separation between your work and home life. If you work from home, it can be harder to find ways to leave work at work, but some strategies include relegating work to one room in the house (and leaving that room at the end of the day!), setting a timer, shutting down the computer, and leaving your house each day for lunch.

    - **Boundaries with fellow counselors, friends, or family members:** When you are in a "helping" profession—where everyone expects you to lend them a compassionate ear—it is important to be able to say no more often to the things that drain you and yes to the things that light you up!

- **Refine your time management skills:** Effective time management is imperative for therapists, as your schedule can change frequently. Use a planner for client appointment and set reminders on your phone for other important meetings or deadlines. This will make your job much easier.

- **Get outdoors and spend time in nature:** Enough said.

With these tips in mind, it's time to create your therapist healing toolkit. After you recognize what type of self-care tools are unique to your well-being, write them in the spaces provided and create a schedule each week that will help you prioritize these tools.

Self-Care Tools for the Mind

Self-Care Tools for the Body

Self-Care Tools for the Soul

Conclusion

Now that you (and your clients!) have created a personalized healing toolkit, it's important to remember that the healing process is never over—and it is rarely linear. Life is unpredictable, and it throws adversity at us from every direction. However, it is your responsibility as a mental health clinician to provide clients with the necessary tools and confidence to heal their anxiety, even when their life feels out of control. The healing toolkit is a way for clients to manage their daily anxieties and build the confidence they need to face life's inevitable challenges head-on. Encourage them to incorporate these tools into their everyday lives so they can better stay inside their window of tolerance when adversity arises.

I also want to emphasize that managing symptoms is not the only goal of healing anxiety. The ultimate goal is to help your clients lead a more balanced and fulfilling life, free from the grip of fear. Encourage your clients to take small steps toward their desired outcomes and to celebrate each milestone along the way. Since healing is a journey, remind them that setbacks are a natural part of that process. By providing ongoing support and guidance, you can help them stay motivated and resilient in the face of adversity. Ultimately, they will learn that healing their anxiety is not about eradicating it completely, but managing the emotions that come along with it.

References

American Psychiatric Association. (2013). *Diagnostic and statistical manual of mental disorders* (5th ed.). https://doi.org/10.1176/appi.books.9780890425596

Appleton, J. (2018). The gut-brain axis: Influence of microbiota on mood and mental health. *Integrative Medicine, 17*(4), 28–32.

Borza, L. (2017). Cognitive-behavioral therapy for generalized anxiety. *Dialogues in Clinical Neuroscience, 19*(2), 203–208. https://doi.org/10.31887/DCNS.2017.19.2/lborza

Butterfield, A. (2021, December 13). Understanding the cycle of anxiety. *The OCD & Anxiety Center.* https://theocdandanxietycenter.com/understanding-the-cycle-of-anxiety

Carey, T. A., & Mullan, R. J. (2004). What is Socratic questioning? *Psychotherapy: Theory, Research, Practice, Training, 41*(3), 217–226. https://doi.org/10.1037/0033-3204.41.3.217

Clark, D. A., Beck, A. T., & Alford, B. A. (1999). *Scientific foundations of cognitive theory and therapy of depression.* John Wiley & Sons Inc.

Clarke, J. (2023, October 25). Polyvagal theory: How our vagus nerve controls responses to our environment. *Verywell Mind.* https://www.verywellmind.com/polyvagal-theory-4588049

Cully, J. A., & Teten, A. L. (2008). *A therapist's guide to brief cognitive behavioral therapy.* Department of Veterans Affairs, South Central Mental Illness Research, Education, and Clinical Center. https://depts.washington.edu/dbpeds/therapists_guide_to_brief_cbtmanual.pdf

Fujiwara, Y., & Okamura, H. (2018). Hearing laughter improves the recovery process of the autonomic nervous system after a stress-loading task: A randomized controlled trial. *BioPsychoSocial Medicine, 12*(22). https://doi.org/10.1186/s13030-018-0141-0

Gawain, S. (2010). *Creative visualization: Use the power of your imagination to create what you want in your life.* New World Library.

Gerritsen, R. J. S., & Band, G. P. H. (2018). Breath of life: The respiratory vagal stimulation model of contemplative activity. *Frontiers in Human Neuroscience, 12*, Article 397. https://doi.org/10.3389/fnhum.2018.00397

Gill, L. (2017, November 25). Understanding and working with the window of tolerance. *Attachment and Trauma Treatment Center for Healing.* https://www.attachment-and-trauma-treatment-centre-for-healing.com/blogs/understanding-and-working-with-the-window-of-tolerance

Hay, L. (2016). *Mirror work: 21 days to heal your life.* Hay House.

Hubbard, L. (2023, July 11). Behind the mask: Managing high-functioning anxiety. *Mayo Clinic Health System.* https://www.mayoclinichealthsystem.org/hometown-health/speaking-of-health/managing-high-functioning-anxiety

Jungmann, M., Vencatachellum, S., Van Ryckeghem, D., & Vögele, C. (2018). Effects of cold stimulation on cardiac-vagal activation in healthy participants: Randomized controlled trial. *JMIR Formative Research, 2*(2), Article e10257. https://doi.org/10.2196/10257

Kai, S., Nagino, K., Ito, T., Oi, R., Nishimura, K., Morita, S., & Yaoi, R. (2016). Effectiveness of moderate intensity interval training as an index of autonomic nervous activity. *Rehabilitation Research and Practice.* Article 6209671. https://doi.org/10.1155/2016/6209671

Lu, W.-A., Chen, G.-Y., & Kuo, C.-D. (2011). Foot reflexology can increase vagal modulation, decrease sympathetic modulation, and lower blood pressure in healthy subjects and patients with coronary artery disease. *Alternative Therapies in Health and Medicine, 17*(4), 8–14.

Masood, N. (2023, September 13). The anxiety cycle: How to break out of it. *CareClinic.* https://careclinic.io/anxiety-cycle

Mayo Clinic. (2018, May 4). *Anxiety disorders.* https://www.mayoclinic.org/diseases-conditions/anxiety/diagnosis-treatment/drc-20350967

McCorry, L. K. (2007). Physiology of the autonomic nervous system. *American Journal of Pharmaceutical Education, 71*(4), Article 78.

Medanta. (2019, April 21). The dark side of anxiety: 7 effects of anxiety on the body. https://www.medanta.org/patient-education-blog/the-dark-side-of-anxiety-7-effects-of-anxiety-on-the-body

Meek, W. (2021, January 25). 5 ways anxiety can be helpful. *Verywell Mind.* https://www.verywellmind.com/top-ways-anxiety-is-helpful-1393079

National Institute of Mental Health. (2023, April). Anxiety disorders. https://www.nimh.nih.gov/health/topics/anxiety-disorders

Northwestern Medicine. (2020, June). The science of anxiety. https://www.nm.org/healthbeat/healthy-tips/emotional-health/the-science-of-anxiety

Novak, S. (2021, April 15). Fight or flight? Why our caveman brains keep getting confused. *Discover Magazine.* https://www.discovermagazine.com/health/fight-or-flight-why-our-caveman-brains-keep-getting-confused

Noyes, R. (2001). Comorbidity in generalized anxiety disorder. *Psychiatric Clinics of North America, 24*(1), 41–55. https://doi.org/10.1016/s0193-953x(05)70205-7

Porges, S. W. (2004). Neuroception: A subconscious system for detecting threats and safety. *Zero to Three, 24*(5), 19–24.

Porges, S. W. (2007). The polyvagal perspective. *Biological Psychology, 74*(2), 116–143. https://doi.org/10.1016/j.biopsycho.2006.06.009

Porges, S. W. (2009). The polyvagal theory: New insights into adaptive reactions of the autonomic nervous system. *Cleveland Clinic Journal of Medicine, 76*(4, Suppl. 2). S86–S90. https://doi.org/10.3949/ccjm.76.s2.17

Risser, M. (2022, January 20). Therapist burnout: Signs, causes & 17 self care strategies. *Choosing Therapy.* https://www.choosingtherapy.com/therapist-burnout

Schenck, L. K. (2011, August 28). 5 steps to full consciousness. *Mindfulness Muse.* https://www.mindfulnessmuse.com/mindfulness/5-steps-to-full-consciousness

Shafir, H. (2022, November 9). Perfectionism: Signs, causes, & ways to overcome. *Choosing Therapy.* https://www.choosingtherapy.com/perfectionism

Siegel, D. J. (1999). *The developing mind: Toward a neurobiology of interpersonal experience.* Guilford Press.

Spitzer, R. L., Kroenke, K., Williams, J. B. W., & Löwe, B. (2006). A brief measure for assessing generalized anxiety disorder. *Archives of Internal Medicine, 166*(10), 1092–1097. https://doi.org/10.1001/archinte.166.10.1092

Strack, J., Lopes, P., Esteves, F., & Fernandez-Berrocal, P. (2017). Must we suffer to succeed?: When anxiety boosts motivation and performance. *Journal of Individual Differences, 38*(2), 113–124. https://doi.org/10.1027/1614-0001/a000228

Weathers, F. W., Litz, B. T., Keane, T. M., Palmieri, P. A., Marx, B. P., & Schnurr, P. P. (2013). The PTSD checklist for *DSM-5* (PCL-5) – LEC-5 and Extended Criterion A [Measurement instrument]. *National Center for PTSD*. https://www.ptsd.va.gov/professional/assessment/adult-sr/ptsd-checklist.asp

Wright, J. H. (2006). Cognitive behavior therapy: Basic principles and recent advances. *FOCUS, 4*(2), 173–178. https://doi.org/10.1176/foc.4.2.173

Zolfagharifard, R. (2023). Subclinical symptoms of mental health problems. *The Licensed Confidant*. https://www.thelicensedconfidant.com/subclinical-symptoms

About the Author

**Alison Seponara, MS, LPC,** is a licensed psychotherapist who specializes in cognitive behavioral therapy and mindfulness-based positive psychology. Known as @theanxietyhealer on Instagram with a following of over 574K, she is dedicated to ending the stigma of mental health and providing a safe, supportive community free of judgment and bias. Alison is also the host of *The Anxiety Chicks* podcast, where she uses her therapeutic expertise and her own life experience to educate listeners on anxiety disorders, the mind-body connection, and holistic remedies—and most importantly, how to keep it real when it comes to talking about mental health. Alison is the founder of The Anxiety Healing School, an online curriculum that helps students learn how to combat intrusive thoughts, face fears, and rewire the anxious brain. She is the author of *The Anxiety Healer's Guide*. She lives and works in Pennsylvania.